GW00802195

Fluid, Electrolytes, Acid-Base and Nutrition

Fluid, Electrolytes, Acid-Base and Nutrition

ARTHUR K.C. LI

MICHAEL R. WILLS

GILLIAN C. HANSON

1980

ACADEMIC PRESS

A Subsidiary of Harcourt Brace Jovanovich, Publishers

LONDON · NEW YORK · TORONTO · SYDNEY
SAN FRANCISCO

ACADEMIC PRESS INC. (LONDON) LTD.
24/28 Oval Road,
London NW1

United States Edition published by
ACADEMIC PRESS INC.
111 Fifth Avenue
New York, New York 10003

Copyright © 1980 by
ACADEMIC PRESS INC. (LONDON) LTD.

All rights Reserved
No part of this book may be reproduced in any form by photostat, microfilm, or any other means, without written permission from the publishers

Li, Arthur K C
Fluid, electrolytes, acid-base and nutrition.
1. Body fluids 2. Acid-base equilibrium
3. Electrolyte metabolism
I. Title II. Wills, Michael Ralph
III. Hanson, Gillian Coysh
612′.01522 QP90.5 80-49984

ISBN 0-12-448150-7

Printed in Great Britain at the Alden Press, Oxford.

Preface

This book is written for senior students and newly qualified doctors as a short, simple guide to the understanding of fluid and electrolyte balance, acid-base homeostasis and nutritional therapy. It seeks to explain the basic principles and the rationale behind the management of these important, and yet often considered "uninspiring", topics. Little bibliography has been given, but it is essential to seek advice and guidance from senior colleagues when faced with complicated situations.

ARTHUR K. C. LI, M.A., M.B., B.Chir., F.R.C.S. (Eng.)
Senior Lecturer and Consultant Surgeon,
Academic Department of Surgery, Royal Free Hospital, London, England

MICHAEL R. WILLS, R.D., M.D., Ph.D., M.R.C.P., F.A.C.P., F.R.C.Path.
Professor of Pathology and Internal Medicine, University of Virginia,
Charlottesville, Virginia, USA

Formerly: *Professor of Pathology and Director of the Metabolic Unit,*
Royal Free Hospital, London, England

GILLIAN C. HANSON, M.B., F.R.C.P.
Consultant Physician in Charge,
Intensive Therapy Unit, Whipps Cross Hospital,
Leytonstone, London, England

Contents

2
Acid-Base and Hydrogen Ion Homeostasis
M.R. WILLS 27

3
Nutrition and Metabolism
G.C. HANSON 55

One

Fluid and Electrolyte Homeostasis

A. K. C. Li
Academic Department of Surgery,
Royal Free Hospital, London, UK

DEFINITION OF UNITS OF MEASUREMENTS

In order to grasp the concepts of the control of fluid and electrolyte balance, it is essential to understand fully the quantitative terms used. Rational therapy can then be planned and calculated from the available data.

Weight in grams (g) or kilograms (kg). Gram = the weight of one cubic centimetre of distilled water at the maximum density (4°C).

Volume in millimetres (ml) or litres (litre). Litre = the capacity represented by a cube whose edge is 10 cm.

Concentrations for *electrolytes* in millimoles per litre (mmol/litre) or milliequivalents per litre (mEq/litre).

Atomic weight is the weight of an atom relative to hydrogen which is assumed to be one unit, e.g. Na = 23, K = 39, Cl = 35.5.

Equivalent weight is the weight of a chemically reacting unit relative to hydrogen which is assumed to be one unit, e.g.

$$\text{equivalent weight} = \frac{\text{atomic weight}}{\text{valency}}$$

A one normal solution in strict chemical terms is one which contains the equivalent weight of a substance in grams in 1 litre. Thus, *chemically* normal saline has the equivalent weight of NaCl in grams in 1 litre:

$$= (23 + 35.5)\ \text{g/litre}$$

$$= 58.5\ \text{g of NaCl/litre}$$

$$= 1\ \text{Eq/litre}$$

$$= 1000\ \text{mEq/litre (mmol/litre)}$$

However, in *clinical practice* the term *normal* is used in a physiological sense; the solution is iso-osmotic with red blood cells. A normal physiological saline solution contains approx. 150 mmol of sodium ions per litre.

In any electrically neutral solution, the total number of positively charged cations (Na^+, K^+ etc.) must *equal* the number of negatively charged anions (Cl^-, HCO_3^- etc.). If these ions are expressed in millimoles the total of each must be equal, 1 mmol of any cation can then replace 1 mmol of any other cation and similarly for anions. (In most body fluids anions and cations *each* total about 150 mmol/litre.)

Concentrations for *non-electrolytes* in the past were always expressed as mg/100 ml though they are now usually reported as millimoles per litre which gives the concentration in terms of molecules (see S.I. units):

$$\text{molar concentration} = \frac{\text{g/litre}}{\text{mol. wt}}$$

$$\text{millimolar concentration (mM)} = \frac{\text{mg/litre}}{\text{mol. wt}}$$

Concentrations for *total solutes* are expressed as osmolality or osmolarity. Each particle of solute in a solution exerts the same osmotic pressure whether it is a molecule or an ion and whether it is either monovalent or multivalent. When the particle concentration is expressed as per unit *weight* of solvent (per kilogram of solvent) it is called *osmolality*. When the particle concentration is expressed as a unit *volume* of solvent (per litre of solvent) it is then called *osmolarity*. For biological solutions osmolality (osmol) is used and the osmolality of a solution which is isotonic with red cells is approx. 300 mosmol/litre. Two examples of solutions isotonic with blood are as follows:

$$\text{Electrolytes mosmol/litre} = \frac{\text{mg/litre} \times \text{no. ions/molecule}}{\text{mol. wt}}$$

$$\text{Isotonic NaCl (mol. wt} = 58.5\text{) 300 mosmol/litre} = \frac{\text{mg/litre} \times 2}{58.5}$$

$$\text{NaCl mg/litre} = \frac{300 \times 58.5}{2}$$

$$= 8800 \text{ mg/litre}$$

$$= 0.88\%$$

$$\text{Non-electrolytes mosmol/litre} = \frac{\text{mg/litre (mmol/litre)}}{\text{mol. wt}}$$

$$\text{Isotonic glucose (mol. wt} = 180\text{) 300 mosmol/litre} = \frac{\text{mg/litre}}{180}$$

$$\text{Glucose mg/litre} = 300 \times 180$$

$$= 54\,000 \text{ mg/litre}$$

$$= 5.4\%$$

S.I. units (International System of Units) have been recommended to standardize the results of measurements in science and medicine. The basic concept is not new and implies that the units used in practice should correspond to their actual activity in biological fluids.

As physiological and pharmacological activities are normally proportional to the concentration of molecules and ions present and not to the mass concentration, the *mole* becomes the basic *reacting unit* and these species are chemically interchangeable. That is, 1 mmol of Cl^{-1} will react with 1 mmol of Na^+ or K^+; this is not true of the weight of these substances. In the S.I. system the concentrations of *all* substances, both electrolyte and non-ionized substances that have a known molecular weight, are expressed in terms of a standard number of specific particles per unit volume which is normally the litre. For monovalent ions 1 mol is numerically the same as 1 equivalent.

For multivalent ions like calcium whose normal plasma concentration has previously been expressed as 10 mg/100 ml, this now becomes 2.5 mmol/litre (at. wt of calcium = 40). For non-electrolytes like glucose (mol. wt = 180) the plasma concentration of 90 mg/100 ml now becomes 5 mmol/litre.

Conversion tables are widely available for those biological substances and compounds encountered in clinical practice.

DISTRIBUTION AND BALANCE OF FLUIDS AND ELECTROLYTES

Distribution

The total body fluid content is distributed between the intracellular and extracellular fluid compartments. Extracellular fluid may be further divided into:

plasma
interstitial fluid
secretions

Volumes

The basis of measurement of body fluid compartment volumes is the dilution principle where a known quantity of substance (whose distribution in the body corresponds with the fluid space to be measured) is given and the dilution which results is measured. Thus the volume into which dilution occurred can be calculated.

Total body water = approx. 60% of body weight (measured by heavy water).

Extracellular fluid = approx. 20% of body weight (measured by inulin).

Plasma fluid = approx. 4.5% body weight (measured by labelled albumin).

Intracellular fluid = total body water − extracellular fluid (including plasma) = approx. 40% body weight.

Composition

The electrolyte composition of fluids in various body compartments are shown in Tables 1 and 2.

Table 1

	Millimoles per litre			
	Na^+	K^+	Cl^-	HCO_3^-
Intracellular fluid	10	100	10	10
Extracellular fluid				
plasma	140	4	100	26
interstitial	130	4	110	26
secretions (see Table 2)				

Balance of Body Fluids

The total input of fluid must equal the total output to maintain balance. It is therefore important to recognize the various routes and the potential amounts of fluid lost in both normal and pathological situations. The mechanisms for volume control are closely linked with those for sodium balance.

The *minimal obligatory* daily output is from:

kidneys as urine	500 ml
skin as sweat	400 ml
respiration as evaporation	500 ml
gastro-intestinal tract as faeces	100 ml
	1500 ml

Table 2 *Electrolyte composition of the body secretions*

	Millimoles per litre					
	H^+	Na^+	K^+	HCO_3^-	Cl^-	Comment
Saliva		20–30	10–20		20–30	
Gastric	40–60	20–80	5–20		100–150	The more acidic the gastric juice, the less sodium ions it contains
Biliary		120–140	5–15	30–50	80–120	
Pancreatic		120–140	5–15	70–110	40–80	
Small bowel		120–140	5–20	20–40	80–120	
Large bowel (diarrhoea)		120–140	20–80	20–40	80–120	
Sweat		60–80	10–20		40–80	The levels are variable and are reduced by adreno-cortical hormones and raised in fibrocytic disease
Cerebrospinal fluid		135–145	3–5	24–28	110–130	c.s.f. chloride usually maintains constant ratio of 1.2 to plasma chloride

Thus the *minimal* daily input of fluids to maintain balance in a non-catabolic patient must be around 1500 ml. This may be made up by water from:

metabolism	500 ml
oral intake	1000 ml
	1500 ml

Normally, in a healthy adult, the daily oral water intake is 2–3 litres and well exceeds the minimal amount which is required to maintain balance. In abnormal situations fluid may be lost in much greater amounts from those four routes, as well as from other routes. Examples of excessive fluid losses include:

from the kidneys	diabetes mellitus diabetes insipidus
from the skin	heat stroke

from respiration	tachypnoea from any cause
from the gastro-intestinal tract	gastric loss in vomiting and aspiration fistulae diarrhoea

In these circumstances oral replacement may not be possible and parenteral infusion (intravenous, intraperitoneal or by the rectal route) becomes essential.

Balance of Electrolytes

The output of electrolytes must be replaced by input to maintain balance. The distribution of electrolytes within the body compartments must be known so that the variations in serum values can be correctly interpreted.

The absorption site of electrolytes is from the gut and the excretory route is predominantly through the kidneys. All electrolyte disorders are commonly associated with either intestinal or renal dysfunction.

Sodium. The normal daily requirement is 100 to 200 mmol and 95% is absorbed from the gastro-intestinal tract. The normal plasma values are 135 to 145 mmol/litre. The distribution is mainly in the extracellular fluid (e.c.f.). Sodium losses from the kidneys are controlled by hormonal output from the adrenal gland, renal blood flow, and osmoreceptors (ADH) cited in the brain. Further losses occur from the skin. Plasma sodium changes tend to reflect the total body sodium since most of the sodium ions are within the extracellular fluid.

Hypernatraemia, however, can indicate not only a total body increase but also dehydration from primary water depletion. Similarly, hyponatraemia may indicate overhydration from water intoxication. Thus, the level of serum sodium is closely linked with water balance. In the "sick cell syndrome" hyponatraemia occurs because the sodium pump becomes ineffective and sodium ions are sequestered outside the intravascular space.

Chloride. Chloride ions are mainly absorbed with sodium from the gastro-intestinal tract and the distribution is approx. 75% in the e.c.f. and 25% in the intracellular fluid (i.c.f.). Normal plasma values are 95 to 105 mmol/litre. Chloride losses are largely from the kidneys by a mechanism which is independent from that for sodium. Some losses however, occur from the skin. The chloride ion is essentially inert and hyperchloraemia effects changes in the concentration of other ions, especially bicarbonate.

Potassium. The normal daily requirement for potassium ions is 60 to 80 mmol of which approx. 90% is absorbed from the gastro-intestinal tract. The normal plasma values are 3.5 to 5 mmol/litre. Potassium is largely distributed in the i.c.f. Potassium loss is largely from the kidneys and the excreted amount is the difference between glomerular filtration and tubular reabsorption. A smaller amount is lost in the faeces. The potassium ion is mainly an intracellular ion and therefore, the plasma value *does not* reflect the state of total body stores.

The potassium concentration of the e.c.f. fluid and plasma is closely associated with acid-base status – hyperkalaemia occurs with acidosis and hypokalaemia with alkalosis.

The plasma potassium value is important as this ion can have serious effects on the myocardium, particularly in patients on digitalis therapy; ventricular fibrillation or cardiac arrest can occur at either high or low concentrations.

Bicarbonate. The distribution is in both e.c.f. and i.c.f. The normal plasma values are 24 to 28 mmol/litre. Bicarbonate is lost from the body in the urine, and from the lungs as carbon dioxide. The bicarbonate ion is central in acid-base homeostasis (see Chapter 2). The steady-state regulation of the concentration of this ion affects both hydrogen and potassium status.

For the balance of water and electrolytes it can be seen that in uncomplicated situations the average daily requirements in *health* are as follows: *water* – 2 to 3 litres as fluid and foodstuff (see p. 7) containing 100 to 200 mmol of *sodium* (see p. 8) and 60 to 80 mmol of *potassium* (see above). These must be the working quantities from which daily replacement therapy is calculated. These values may need to be modified in the light of the results of the investigations undertaken for each individual patient, especially in the presence of either complication and/or stress.

DIAGNOSIS OF FLUID AND ELECTROLYTE IMBALANCE

Patient Evaluation

The clinical diagnosis of fluid and electrolyte disturbances is critically dependent on the awareness of the type of patient who is at risk. Such patients include those with gastro-intestinal disorders, including those with fistulae and/or gastric suction, renal diseases, including long-term diuretic therapy, hepatic disturbances, and any patient on intravenous therapy. Evaluation of the patient suspected to be suffering from fluid and/or electrolyte imbalance is as follows.

History

From the outset all fluid losses and replacements must be carefully recorded. This may involve catheterization of the patient in order to collect accurately the total urine output. An assessment should also be made of the fluid losses as sweat or through ventilation. From these data, the total volume of fluid lost will enable water balance to be maintained, while the nature of the fluid loss will reflect not only the electrolyte status but also the replacement requirements.

Thirst is a useful but *not absolute* sympton of water depletion and hypernatraemia. This symptom may be absent on an old, "ill" or confused patient; or present with hypercalcaemia, or due to a "dry mouth" from mouth breathing.

Clinical Examination

Skin and subcutaneous tissue. The state of the skin and subcutaneous tissue (by pinching a fold of skin) is not an entirely reliable method for estimation of body fluid status. Dehydration of greater than 4% of the body weight is usually detectable by the presence of dry inelastic skin – but this may also apply in severe malnutrition in the absence of fluid depletion. The presence of detectable systemic oedema usually represents fluid excess of approx. 6% of the body weight.

Blood pressure – peripheral perfusion. When the blood pressure falls for any reason there is generally compensatory peripheral vasoconstriction and a rising pulse rate. The pulse rate rises at an early stage when the blood volume decreases but the blood volume may fall by one-third before the systolic blood pressure starts to decrease. In healthy young adults the vasoconstriction may be so marked as to give a mild hypertension at the initial phase of a falling blood volume. In the absence of drugs producing a bradycardia (e.g. β blockers) a tachycardia is invariably present.

It is important to realize that hypotension may arise from factors other than a reduced blood volume. These factors must be excluded, e.g. cardiogenic and septic shock.

Peripheral vasoconstriction due to a falling circulating blood volume may be prevented should the patient be receiving treatment which prevents vasoconstriction – this includes spinal or epidural anaesthesia (where vasodilatation occurs below the block) and the use of certain hypotensive agents.

Urine output. A falling urine output (less than 40 ml/h) in the presence of concentrated urine (osmolality greater than 800 mosmol/kg) is indicative of impaired renal perfusion and is commonly related to a falling circulating blood volume. This level may fall to 400 mosmol/kg during and after anaesthesia and for 24–36 h after surgery.

Respiratory state. Hyperventilation may be related to pulmonary oedema or be compensatory to a metabolic acidosis. Pleural effusions may be present in hypoalbuminaemia or chronic fluid overload. It is important that an intrinsic pulmonary cause should be excluded.

The abdomen. The anterior abdominal wall may show varices or be oedematous. Bruising in the flanks or around the umbilicus suggests pancreatitis. Free fluid in the peritoneal cavity is commonly related to liver failure and hypoalbuminaemia but may also be present following a perforation, pancreatitis, neoplasm or infective peritonitis. Paralytic ileus commonly develops following intra-abdominal sepsis or recent bowel surgery, but in the absence of these factors may suggest hypokalaemia or severe generalized illness, e.g. septicaemia, hypothyroidism.

Investigations

The entire blood should be looked at as a whole and in relationship to the history and patient examination and fluid balance. Although a single result is useful it is preferable to follow the pattern of a series of results, especially in a patient on treatment.

Haemoglobin and haemotocrit. The estimations are useful indices of haemoconcentration, provided that no blood loss has occurred; pre-existing anaemia or polycythaemia will affect these two variables.

Urea. The plasma concentration is influenced by previous dietary protein intake, hepatic synthesis and renal excretory functions.

If all these processes are normal, then a high value *may* reflect haemoconcentration due to dehydration and water loss. It is important to recognize that a raised value may be seen following a heavy protein meal (or blood loss into the gastro-intestinal tract), or in impairment of renal function. A low value may occur in starvation such as a patient on clear intravenous fluids and gastric aspiration, and in pregnancy where there is increased renal perfusion and expansion of blood volume. The urea concentration is therefore a very non-specific indicator of water balance and total body water status.

Sodium. Since this ion is closely related to water balance, its concentration provides a good indication of the relative state of hydration provided that there is no abnormality in the distribution of fluid between the extracellular and intracellular fluid compartments. Thus as a *generalization*:

hypernatraemia = dehydration

hyponatraemia = relative overhydration

Potassium. 70% of the total body potassium resides in the muscle cell and therefore the serum potassium is a poor reflector of the total body potassium. Alterations in either intracellular or extracellular potassium concentrations affect membrane excitability by affecting membrane potential. This has far reaching effects, in particular on cardiac function and muscle function. Potassium ions readily shift from the intravascular space into the cells. The body constantly attempts to maintain a normal intravascular pH and in order to do this, potassium shifts may occur. In an acidosis, hydrogen ions enter into the cells and are dealt with by intracellular buffers. As hydrogen is a cation, in order to maintain electrical neutrality across the cell membrane the cation potassium leaves the cell; thus hyperkalaemia is associated with a metabolic acidosis; similarly hypokalaemia with a metabolic alkalosis.

In the presence of a metabolic acidosis, if the plasma potassium concentration is either normal or low, it may reflect total body potassium depletion.

Bicarbonate. This plasma estimation should be simultaneously done with potassium as it is a reflector of acid-base balance (see Acid-Base Section).

Chloride. As a generalization hyperchloraemia occurrs in acidosis, while hypochloraemia usually follows excessive gastric loss either due to vomiting or aspiration.

Urinalysis

Biochemical examination of the urine is a valuable assessment of renal function since renal function is involved in normal body homeostasis.

Specific gravity. This is an index of dehydration provided that renal function is normal and that there are no abnormal urine constituents present (for example, glycosuria) which would make this measurement unreliable.

Osmolality. This reflects renal tubular function and specifically the handling of water and renal clearance of sodium ions.

The ratio of plasma osmolality to urine osmolality is a delicate reflector of renal functon provided diuretics have not been used prior to the test. A ratio of 1.2 or less indicates acute intrinsic renal failure and a ratio of 2.7–4 indicates normal renal function. High ratios are seen in severe dehydration.

Sodium and potassium. The 24 h urinary estimations indicate the renal handling of these ions and should be taken into account when calculating electrolyte replacements in a difficult metabolic problem.

METABOLIC RESPONSE TO TRAUMA

The major metabolic responses to trauma consist of the following:

Negative Nitrogen Balance. There is breakdown of protein which is mediated by an increase in catabolic hormones (cortisol, adrenaline and ACTH).

Loss of Potassium. There is disproportionate loss of potassium relative to nitrogen breakdown.

Retention of Sodium. The retention of sodium leads to retention of water; this is mediated by an increase in the secretion of both ADH and aldosterone.

The extent and duration of the metabolic response to trauma depends on the severity of the injury. The changes usually occur immediately and last for approx. 24 to 48 h, although in severe trauma, changes may be of longer duration, albeit to a lesser degree.

MANAGEMENT OF FLUID AND ELECTROLYTE BALANCE

Fluid Replacement

The electrolyte composition of solutions commonly used in intravenous therapy are enumerated in Table 3.

Table 3 *Electrolyte composition of commonly used solutions*

	Millimoles per litre					
	Na^+	K^+	Ca^+	HCO_3^- equiv.	Cl^-	Calories per litre
Normal saline	154				154	
5% Dextrose						200
4.3% Dextrose saline (1/5 Normal saline)	31				31	173
Hartmann's (Ringer-Lactate)[a]	127	5	4	22	112	
Darrow's (Potassium Lactate)[a]	122	36		55	104	
Sodium bicarbonate 1.4%	167			167		
Sodium bicarbonate 8.4%[b]	1000			1000		

[a] *These solutions are not recommended* in electrolyte management because, as can be seen from their composition, they are multiple electrolyte solutions and their use prevents the development of a rational approach.
[b] 8.4% sodium bicarbonate *is hypertonic* and should only be used with extreme caution in correcting the severe acidosis from cardiac arrest.

Blood. Blood may be used for direct replacement of blood loss and for expansion of the vascular compartment, thereby maintaining blood pressure and renal blood flow. The major disadvantages are that it is not immediately available (because of cross-matching), that there are hazards associated with transfusion, and that if not carefully monitored, volume overload may occur (this risk may be minimized by right atrial pressure monitoring (see p. 16)). It should be remembered that unless blood is fresh it does not contain labile clotting factors.

Plasma. Plasma has no delay in expansion of the vascular compartment and fresh frozen plasma contains clotting factors. The disadvantages are: the hazards of transfusion, especially in infection; red cell losses are not replaced; and "overtransfusion" can occur. In using plasma, the therapeutic aim in the treatment of a shocked patient should be to achieve a haematocrit value of 30%; this is in order to prevent an increased blood viscosity and also achieve the optimum oxygen carrying capacity to the cell.

Dextran. As a plasma substitute it contains large molecules, which act as plasma proteins. It is a simple method with which to expand the vascular compartment and improve blood flow. The disadvantages are that it interferes with blood cross-matching as the molecules coat the patient's red cells and, therefore, it should *not* be given until blood has been taken for cross-matching. It can induce allergy by acutely releasing histamine, which may result in severe bronchospasm *without* associated weal formation. Extreme caution must, therefore, be exercised in patients with airway obstruction. As there is evidence suggesting that dextrans may produce a coagulation defect, it is wise not to infuse more than the recommended dose of 1.5 g/kg of body weight/24 h. Patients in shock should not be given Dextran 40 as it may precipitate renal failure.

Normal saline. One litre contains the normal daily requirements of sodium (150 mmol/litre). It can replace all forms of extracellular fluid loss (e.g. vomiting and diarrhoea). The disadvantages are that it does not stay in the vascular compartment but diffuses throughout the e.c.f. and is, therefore, useless in the prolonged maintenance of blood pressure. Furthermore sodium overload can occur.

5% Dextrose. This is an isotonic solution valuable for water replacement. It does not stay in either the vascular or e.c.f. compartment but diffuses throughout the body water store and therefore overhydration is a risk. Moreover, it does not replace any electrolytes.

4.3% Dextrose saline. This is a compromise between normal saline and 5% dextrose. It contains 1/5 of the sodium content of normal saline (30 mmol/litre sodium) and the isotonicity is made up by the content of 4.3% dextrose. With this solution, essentially water is being given together with a small amount of electrolytes.

Hartmann's. This is regarded as a physiological replacement solution since it contains approximately the normal ionic constituents of the e.c.f. compartment. Its value is debatable since in complicated situations, the pre-existing constituent ions make the calculation of replacement therapy complex and unnecessarily difficult, especially if additions of extra ions are necessary.

Route of Fluid Administration

Oral

Fluid and electrolyte balance can commonly be controlled by oral, or Ryles tube, administration. This is the preferable route but should not be relied upon in patients with gastro-intestinal atony, or those who have excessive fluid and electrolyte losses, and in situations of acute and severe electrolyte and fluid imbalance. The oral route should also be avoided in patients who are unable to maintain their airway since there is a danger of pulmonary aspiration. Many critically ill patients do not absorb adequately via the oral route.

In certain situations absorption via the gastro-intestinal tract is possible to a limited degree, this being supplemented by intravenous therapy. Strong tea and fresh juice contain significant amounts of potassium whilst beef extracts contain sodium.

Intravenous Fluid Administration

This can be given by peripheral or central veins.

Peripheral veins. These can tolerate isotonic fluids but this route is unsuitable for hypertonic solutions and for long-term intravenous nutrition as painful thrombophlebitis usually develops within 24 h of administration.

Central veins. These can accept all fluids, and by siting the catheter tip in the right atrium, simultaneous measurement of central venous pressure can be undertaken. A major disadvantage of this route is that if infection supervenes, the outcome can be fatal. Leg veins should not be used for intravenous administration since it may be complicated by deep vein thrombosis and consequent fatal pulmonary embolism.

In situations where the volume of fluid lost has been severe or where there is doubt about patient tolerance to a fluid load, the volume required can be estimated according to sequential readings of the central venous pressure (c.v.p.). (Central venous pressure should be measured in the mid-axillary line with the patient lying horizontally.) In severe hypovolaemia the c.v.p. is generally 2 cm H_2O or less, and the appropriate fluid should be titrated intravenously until the c.v.p. is 4–8 cm H_2O. Where the c.v.p. is 8 cm H_2O or more before commencing the infusion and fluid depletion is still suspected, it is wise to seek a senior opinion.

Fluid administrations via the rectum and intraperitoneal routes are unreliable and should not be used except under exceptional circumstances such as when an intravenous infusion cannot be erected. Water only should be given as the amount of electrolytes that can be absorbed from the colonic mucosa is uncertain, and it should be administered under low pressure by a slow drip (1 litre/12 h).

Both fluids and electrolytes can be absorbed from the peritoneal cavity. The free exchange of electrolytes and water across the peritoneal membrane is used in the treatment of renal failure by peritoneal dialysis.

Drugs Which Can Influence Fluid and Electrolyte Balance

Ion Exchange Resins

These are a group of insoluble acids or bases which can combine with cations (H^+, Na^+, Ca^{++}, NH_4^+) or anions (CH^-, Cl^-) respectively in an exchange reaction. The exchange depends on the natural affinity of the resin for the ions concerned and the concentration of individual ions. These resins are given orally (or occasionally rectally) and are excreted in the faeces where they are in equilibrium with the concentration of faecal electrolytes.

Cation exchange resins. To promote potassium ion exchange, the resin is given in the sodium phase. Following ion exchange in the intestine, potassium is lost in faeces while the exchanged sodium is absorbed. The therapeutic effect of orally administered resin may take up to 24 h; a quicker effect can be obtained by rectal administration.

Anion exchange resins. These are usually given in the hydroxyl form to exchange for chloride in the stomach, thereby reducing acidity. Weak basic resins release chloride further along the intestine and do not actually remove this ion from the body, while strong basic resins are finally excreted in the chloride form.

Diuretics

These may be regarded as kidney "poisons" in that they inhibit normal renal tubular function and thereby cause a diuresis.

Powerful diuretics like *frusemide* have multiple sites of action along the nephron inhibiting sodium reabsorption. As sodium handling in the tubules is closely linked with water transport, a failure of sodium reabsorption results in a water diuresis.

Carbonic anhydrase inhibitors such as acetozolamide (Diamox) act on the renal tubule where they reduce the availability of hydrogen ions in the exchange mechanism for sodium reabsorption. The sodium reabsorption involves an exchange for potassium ions. The overall effect of these diuretics is the passage of an alkaline urine with a high sodium and potassium content. *Aldosterone antagonists* (spironolactones) act on the distal renal tubules where exchange of sodium for potassium takes place under the influence of aldosterone. These compounds produce a mild diuresis with sodium loss and potassium retention. Osmotic diuretics like *mannitol* act by increasing the osmolality of the fluid (urine) within the lumen of the collecting tubules and diminish or reduce the effect of ADH on water reabsorption.

Recommended Regimens for Fluid and Electrolyte Administration

Some general guidelines may be given, but each patient must be treated on an individual basis. Careful monitoring of total fluid and electrolyte administration and measurement of plasma electrolyte concentration are necessary; serum osmolality and acid-base state should be included if indicated. Should there be excessive fluid losses from the sites (e.g. the bowel) the volume and the electrolyte content of the loss should be estimated.

In the establishment of individual patient regimens one must differentiate between requirement and replacement. Normally an adult *requires* approx. 2 litres of water a day to maintain fluid balance (see p. 7). Where there is clinical evidence of fluid depletion *replacement* will be necessary. Replacement should be based upon the estimated losses and with the addition of the estimated volume that is likely to be lost over the subsequent 12 h. To this must be added the maintenance requirement. In situations where evaluation of estimated losses are difficult or where volume tolerance is uncertain, volume should be replaced according to sequential readings of the c.v.p. In less serious situations fluid volume can generally be replaced safely by sequential observations of blood pressure, pulse rate and volume of urine output. It is essential when replacing fluid volume to estimate whether the depletion is predominantly water or a mixed depletion of sodium and water (pp. 20 and 21). When planning intravenous fluid replacement it is important to give saline, blood or plasma, if possible, during the day rather than the night when the renal ability to excrete any overload is decreased as a consequence of the normal circadian renal variation in water and electrolyte excretion.

There is a daily need of between 70 and 150 mmol of sodium ions to maintain balance. One litre of normal saline contains 150 mmol of sodium. 750–1000 ml a day of normal saline is therefore needed for *requirement* and more is needed for replacement if there is a sodium deficit. During periods of stress the requirements for sodium ions are considerably diminished (see below and p. 13) and extreme caution must be exercised to avoid sodium overload with its associated increase in intravascular volume.

Potassium administration should *not* be undertaken until the patient has passed urine of sufficient quantity and/or quality to exclude the possibility that he is suffering from acute renal failure.

In patients on intravenous fluids alone who are not losing excessive volumes of fluid, potassium replacement is rarely necessary in the first 2 days since there is a large intracellular store of potassium, which makes depletion unlikely. Potassium replacement may be required earlier should the patient be receiving thiazide diuretics, steroids or intravenous feeding, or if the patient has glycosuria.

In an oliguric patient with a plasma potassium concentration of 3.2 mmol or less the advice of a senior colleague should be obtained since on rare occasions the administration of potassium may be indicated. On rare occasions oliguria or anuria may develop as a direct consequence of severe hypokalaemia.

Should potassium be given before renal function has been shown to be adequate, and there is an associated metabolic acidosis, dangerous hyperkalaemia may ensue.

Normally 60 to 80 mmol of potassium ions should be given daily; this should be spread over the 24 h period by adding them to the intravenous fluids. Undiluted potassium salts must *not* be given directly into a vein.

In certain clinical situations there may arise problems which require specific attention. In patients with chronic liver disease sodium is commonly retained and poorly excreted. The daily sodium requirement is therefore negligible and no replacement is needed. Fluid should be given as 5% dextrose in an amount which is sufficient to replace the insensible water loss.

The normal metabolic response to stress is sodium retention, this lasts on average 48 h after the stress phase is over. Sodium ions should not be given during this period unless there is evidence of sodium loss (e.g. gastro-intestinal aspirate).

In patients with paralytic ileus, salt containing fluid is sequestered in the bowel resulting in sodium loss. Under such circumstances fluid should be replaced as normal saline.

Volume replacement can be monitored by means of titration against a central venous pressure in order to achieve a normal value of + 4 to + 8 cm of water (see p.16). *Electrolyte replacement* can be gauged by collecting and analysing the volume and electrolyte content of all external fluids lost (e.g. via Ryles tube or from fistula site) and by sequential estimation of plasma values.

CLINICAL DISORDERS

Disorders of Water and Salt Metabolism

Dehydration

This may be defined as a diminished extracellular fluid volume, and may be due to predominant water depletion, predominant sodium depletion, or a combination of both.

Predominant water depletion. This may arise from insufficient intake (nausea, vomiting, coma, physical disability, inability to swallow), excessive loss from osmotic diuresis (e.g. diabetes mellitus), or from failure of tubular re-absorption of water, as in diabetes insipidus.

The consequence of predominant water loss is a *rise in plasma sodium concentration* and a shift of water from the extravascular to the intravascular compartments. The patient, if conscious, complains of severe thirst, and because the intravascular volume is conserved until late, hypotension (unlike in predominant sodium depletion) does not develop until late. In severe depletion, the patient is drowsy and commonly pyrexial. The hypernatraemia causes an increase in anti-diuretic hormone (ADH) secretion and thus a small amount of concentratated urine with a high sodium content is produced.

Treatment. This consists of replacement of water by mouth or by the Ryles tube or in more serious situations with 5% dextrose intravenously. The volume of fluid infused should be gauged by the patient's mental state, blood pressure, pulse rate and urine output. Care must be taken not to swing from water depletion to overhydration, central venous pressure monitoring is therefore advisable in severe cases.

Predominant sodium depletion. This will result in failure of maintenance of the extracellular fluid volume. Symptoms and signs are largely related to a decrease in intravascular volume.

The condition may arise from inadequate replacement of sodium ions in the presence of a dominant sodium loss, e.g. excessive loss of gastro-intestinal secretions, or from urinary sodium loss as in chronic renal disease or adrenal failure (Addison's disease).

On occasions there may be excessive loss of sodium from the skin surface as in sweat, e.g. fibrocystic disease, or following skin loss as in burns.

The clinical picture varies with the time taken to develop depletion and the magnitude of the loss. Mild depletion (sodium deficit approx. 8 mmol/kg) produces weakness and postural fainting. Moderate to severe depletion (8–12 mmol/kg body weight) produces hypotension and oliguria.

A low serum sodium does not necessarily imply a fall in total body sodium, the other causes being dilutional as in water intoxication or a shift of water into the intravascular space related to an osmotic load (e.g. glucose or mannitol). Sodium ions may also move out of the intravascular space in hypokalaemia or the "sick cell" syndrome (see p. 8).

Treatment. The estimation of sodium replacement based on the serum sodium concentration should only be undertaken when there is clear evidence of a sodium loss. Water replacement *per se* in a patient suffering from predominant sodium depletion will cause a further fall in serum sodium.

Treatment in moderate to severe depletion will require intravenous normal saline. A guide to the amount of sodium lost in a patient with dominant sodium depletion can be assessed according to the following formula:

$$\text{body weight in kg prior to illness} = W$$

$$\text{e.c.f. volume in litres} = F$$

$$\text{e.c.f. volume } (F) = W \times \frac{15}{100}$$

$$\text{sodium deficit mmol/litre} = \text{normal serum sodium} - \text{observed serum sodium}$$

$$\text{total sodium deficit} = F \times \text{sodium deficit (in mmol)}$$

The sodium deficit should normally be replaced with normal saline. In cases of severe depletion associated with hypotension, senior advice must be obtained.

Mixed sodium and water depletion. This usually results from losses in the gastro-intestinal tract.

Treatment. This consists of replacement with normal saline and 5% dextrose.

Overhydration

This may be due to a dominant excess of water (water intoxication) or an excess of sodium and water.

Water intoxication generally arises as a result of a combination of factors. These include excessive administration of water either by mouth via the Ryles tube or intravenously; impaired water excretion due to acute intrinsic renal failure, or secondary to high circulating ADH levels (e.g. following stress); or following the use of drugs which impair water secretion, e.g. steriods, syntocinon.

Water intoxication may also follow irrigation of a hollow viscus with water, e.g. the bladder or colon.

Water excess results in cellular overhydration, hyponatraemia and eventually cerebral oedema and death.

Neurological changes, semicoma, irritability or convulsions indicate that serious cerebral oedema has developed and senior advice must be obtained. Under such circumstances radical measures may have to be instituted, e.g. intubation and hyperventilation or administration intravenously of twice normal saline, in order to prevent coning.

In less serious situations water must be restricted and where renal function is adequate, elimination may be accelerated by the use of a thiazide diuretic.

Water and sodium excess. This results in an increase in extracellular fluid volume and is clinically recognized as oedema. A raised jugulo-venous pressure and pulmonary oedema may also be present.

This condition may follow infusion of excessive volumes of salt and water in the presence of inadequate renal clearance, e.g. following stress, hepatic failure, cardiac failure, the use of salt and water retaining drugs, e.g. steroids and from failure to restrict fluids in the presence of inadequte water and sodium clearance.

Treatment. In situations where there is inadequate water clearance as in renal failure, renal dialysis should be considered. Treatment depends on the cause and usually both sodium restriction and diuretics are necessary. Daily weighing of the patient provides a good guide to the response to treatment.

Disorders of Potassium Metabolism

Potassium Depletion

This may occasionally follow a diminished intake, as in malnutrition and anorexia, or follow defective absorption of potassium as in steatorrhoea. Excessive potassium loss can occur from the gastro-intestinal tract (diarrhoea, vomiting, fistulae) or by excessive losses in the urine as may occur during the diuretic phase of acute renal failure and following relief of obstruction in an obstructive uropathy.

Excessive quantities of potassium may also be lost in the urine during therapy with steroids or diuretics and in certain endocrinological disorders (primary hyperaldosteronism and Cushing's syndrome). Under certain conditions the total body potassium may be depleted but the serum potassium may remain normal until late on in the disease, e.g. hypercatabolic illness and hyperglycaemia.

Depletion of total body potassium may or may not be associated with symptoms, this depending upon the rate of development and the ratio of intracellular to extracellular potassium.

Symptoms include muscle weakness with other vague complaints, such as tiredness and depression. Tachycardia may be a manifestation of potassium depletion especially in a patient who has been digitalized; the tachycardia may be associated with a variety of conduction defects. The e.c.g. characteristically shows flattening of the T wave leading to eventual inversion, and widening of the QRS complex with the appearance of a U wave.

Biochemically the serum potassium is less than 3.4 mmol/litre and this is commonly associated with a metabolic alkalosis. Treatment depends upon the severity of the hypokalaemia, the rapidity of onset and whether potassium loss is continuing. In hypokalaemia of slow onset, where excessive losses are not continuing and where the serum potassium is 3 mmol/litre or more, oral therapy is preferable provided absorption is reliable. In chronic severe depletion (serum potassium less than 3 mmol/litre) potassium should be infused as potassium chloride in 1/5 normal saline at a rate of approx. 20 mmol every 3 h in concentration of 20 mmol/250–500 ml of fluid. The patient should be connected to an e.c.g. monitor during the infusion.

In acute moderate potassium depletion (loss of approx. 10% of the total body potassium), the serum potassium is generally 3 mmol/litre or more. One-third to a half of the estimated potassium loss should be replaced over the first 24 h and to this must be added any continuing losses (e.g. gastro-intestinal).

Acute severe potassium depletion (greater than 10% loss of total body potassium) may occur rapidly in the presence of heavy gastro-intestinal losses. Once the serum potassium is 2.8 mmol/litre or less, there is an extreme danger of serious cardiac arrhythmias developing. Under such conditions dextrose solutions must not be used and should there be a metabolic acidosis, this should not be corrected since these measures may precipitate a hypokalaemic crisis. Such patients should be treated as emergencies, and should be transferred to an Intensive Therapy Unit where potassium is infused under e.c.g. monitoring control. Senior advice regarding the rate and concentration of the potassium infusion is essential.

A cardiac arrest from hypokalaemia complicated by a ventricular dysrhythmia should be treated with external cardiac massage, intubation, hand ventilation on oxygen and rapid infusion of approximately 30 mmol potassium chloride in 200 ml 1/5 normal saline over a period of 20 minutes. D.C. cardioversion can be used but in the absence of simultaneous potassium infusion may be complicated by asystole. Rapid infusion of potassium may be complicated by a short-lived period of tetany.

It is essential to check serum magnesium, calcium and inorganic phosphate, since they are commonly low in the presence of severe hypokalaemia.

Potassium Intoxication

This may follow renal failure, where there is impaired potassium clearance, metabolic and/or respiratory acidosis, and adrenal insufficiency. Increased cellular breakdown, as in hypercatabolic states or hyperthermia and acute intravascular haemolysis, may also produce a serious rise in serum potassium. Iatrogenic causes include potassium sparing diuretics, excessive potassium administration, massive transfusion and the use of the muscle relaxant, succinyl choline.

The diagnosis may be suspected when one of the preceding factors

Table 4 *Emergency management of acute hyperkalaemia*

Drug	Dose	Onset of action	Duration of action	Comments
Calcium gluconate Calcium chloride	10% 0.25 mmol Ca^{++}/ml 30 ml i.v. over 3 min 10% 0.45 mmol Ca^{++}/ml 15–20 ml i.v. over 5 min	Min	Less than 1 h	Does not alter the serum potassium; rapidly eliminates cardiotoxicity; calcium chloride necrotic to veins
Sodium bicarbonate	8.4% 1 mmol HCO_3^-/ml Infuse 80–120 ml over 20 min	Min	Up to 2 h when given as a stat dose	Danger of sodium overload. May be infused continuously 80–120 mmol/h. Method of choice when there is hyponatraemia and hypovolaemia
Insulin and dextrose	40% dextrose 4 g CHO/ml Infuse 20–50 g with insulin 1 unit/4 g dextrose over 15–20 min	Min	Up to 4 h	Danger of hyperglycaemia. May be infused continuously 6-hourly with appropriate insulin dosage
Exchange resin in sodium or calcium phase	50–60 g in a sorbitol retention enema or via Ryles tube	approx. 30 min	4–6 h	Sodium overload with sodium phase resin. Diarrhoea. Has to be given via the gastro-intestinal tract

are present and the patient complains of burning and numbness of the extremities, or there is semicoma and clinical evidence of an ascending paralysis. The diagnosis is confirmed by the serum level and the e.c.g. which shows progressive widening of the QRS and peaking of the T wave. Peaking of the T wave generally occurs when the serum potassium is 5.5 mmol/litre or above, though T wave peaking is not an invariable feature.

Treatment. When the plasma potassium concentration is 7 mmol/litre or more, treatment should be instituted immediately. Methods available are enumerated in Table 4. Should renal failure be present, dextrose and insulin are generally used followed by renal dialysis. In non-catabolic renal failure, dextrose and insulin followed by an exchange resin may be considered so that dialysis may be deferred.

In acute hyperkalaemia where renal potassium clearance is normal, the insulin dextrose regimen is generally used. Should there be evidence of cardiotoxicity, calcium chloride should be given initially and a limited quantity of sodium bicarbonate (50–100 mmol) may be considered if there is a severe metabolic acidosis (pH 7.2 or less).

If the cause for hyperkalaemia cannot be readily elicited, senior advice must be obtained.

Two

Acid-Base and Hydrogen Ion Homeostasis

M. R. Wills

Department of Pathology, University of Virginia, Charlottesville, Virginia, USA

In order to understand acid-base and hydrogen ion homeostasis in health and thus disturbances in disease states, it is essential to have a clear concept of the terms used and the physiological control mechanisms involved.

DEFINITIONS

Acid and Base

During the past, these terms have been used to denote several different concepts, but nowadays the definitions used are those which were originally proposed by Brönsted and Lowry.

The definitions are expressed by the equation:

$$\text{Acid}^{n} \rightleftharpoons H^{+} + \text{Base}^{n-1}$$

An *acid* is a molecule which will give off a hydrogen ion under the given conditions, it is a *proton donor.*

A *base* is a molecule which will accept a hydrogen ion under the given conditions, it is a *proton acceptor.*

A molecule which will neither donate nor accept hydrogen ions under the given conditions is an *aprote.*

The more readily an acid gives up a hydrogen ion the stronger it is and similarly the more avidly a base takes up a hydrogen ion the stronger it is.

In the reaction:

$$HB \rightleftharpoons H^{+} + B'$$

A hydrogen ion is given up by the acid HB; the reaction is reversible so that B', the anion left after the loss of the hydrogen ion, is by definition a base and is termed the *conjugate base* of the acid HB; HB is the *conjugate acid* of the base B'; the acid HB and its conjugate base B' are termed a *conjugate pair.*

Acid-base reactions can be described by the equation:

$$A_1 + B_2 \rightleftharpoons A_2 + B_1$$

where A_1 and B_1 represent one conjugate pair and A_2 and B_2 represent a second conjugate acid-base pair. Thus for acetic acid in water:

$$CH_3COOH + H_2O \rightleftharpoons H_3O^{+} + CH_3COO'$$

where CH_3COOH and CH_3OOO' represent one conjugate acid-base pair and H_3O^+ and H_2O represent the second conjugate pair.

The Brönsted–Lowry concept takes into account the central role of water in acid-base reactions in aqueous systems. Water appears in acid-base pairs twice:

$$2H_2O \rightleftharpoons H_3O^+ + OH'$$

In this concept water is simultaneously *a very weak base* and also a *very weak acid*. (In the former situation it takes up a hydrogen ion to form the *strong acid* OH_3^+, while in the latter it gives up a hydrogen ion to form the *strong conjugate base* OH'.)

In all of these definitions it must be stressed that the all important common feature is the *hydrogen ion*. The key to the development of an understanding of acid-base homeostasis in health, and thus disorders and therapy in disease states, hinges entirely around this ion.

pH

pH is strictly defined as the negative logarithm of the molal hydrogen ion activity, thus pH bears a theoretically simple relationship to hydrogen ion concentration $[H^+]$:

$$pH = \log\frac{1}{[H^+]}$$

Buffers

A buffer may be defined as a substance which by its presence in a solution increases the amount of acid or alkali that must be added to cause a unit change in pH.

Weak acids with their conjugate bases act as buffers. In biological fluids these acids include carbonic, phosphoric and other organic acids and the acid (hydrogen ion donor) groups of protein molecules. The haemoglobin in the erythrocytes plays a major role in the buffering of the plasma compartment.

In *plasma*, the principal buffer system is carbonic acid and sodium bicarbonate, and the pH of the plasma compartment is always proportional to the ratio of these two substances irrespective of their absolute concentrations. At a pH of 7.40 the ratio of bicarbonate to carbonic acid is 20:1 and any change in pH must be accompanied by a change in this ratio.

All buffers are in equilibrium with each other and with a change in pH all of them will show a change in this ratio of conjugate base to acid which is necessarily the same change.

The buffering in plasma is more efficient for acids than alkalis but this is not true of all the buffer systems. In practice, the pH change from such additions would be smaller because of the other buffers in the plasma compartment and within cells, and also because of respiratory control mechanisms.

The most effective buffering ability of a buffer-pair system is within a range of ± 1 pH units of its pK value, and is nil at pH values more than 1.5 units from the pK. The pK is equivalent to approximately the pH at which dissociation of the buffer-pair is complete. The bicarbonate carbonic-acid system is the principal buffer-pair of the extracellular fluid compartment but with a pK of 6.1 this buffer-pair system is far from its maximum effect, and is limited both in its buffering capacity and buffering potential. The importance, however, of this buffer-pair system in the extracellular compartment is due mainly to the unique nature of the bicarbonate ion. This "decomposes" on combination with a hydrogen ion liberating CO_2 which, being freely diffusible, can be rapidly removed by the lungs. In this respect the extracellular fluid compartment may be regarded as an "*open*" system because of the freely diffusible nature of carbon dioxide. The initial buffering of a metabolically generated hydrogen ion in the extracellular fluid compartment by the bicarbonate/carbonic acid system involves the simultaneous loss of bicarbonate ion and generation of carbon dioxide. The ultimate excretion of that metabolically generated hydrogen ion from the body is undertaken in the *distal* renal tubule with the simultaneous replacement of the bicarbonate ion lost in the initial buffer action. The bicarbonate/carbonic-acid buffer system, and thus the extracellular hydrogen ion concentration, is therefore maintained in a steady state by the respiratory and renal control of that system. It is, however, important to recognize that the final route of hydrogen ion excretion from the body is to all intents and purposes limited to the kidneys.

Carbon Dioxide System

Bicarbonate ion concentration. The strict definition is the concentration of the HCO_3' ion in a biological fluid. In most physiological studies the bicarbonate ion concentration is calculated as the total carbon dioxide concentration minus the dissolved CO_2 concentration and thus in a physiological sense the term includes carbamino compounds and carbonate plus bicarbonate.

Partial pressure of carbon dioxide (P_{CO_2}). The P_{CO_2} of a biological fluid is by definition the partial pressure of carbon dioxide in a gas phase which is equilibrium with the biological fluid.

Dissolved CO_2 concentration. This is strictly the concentration of the physically dissolved CO_2 gas in a biological fluid; however, H_2CO_3 is usually included. The sum is designated as $S \times P_{CO_2}$ where S is the coefficient relating to the sum of the concentrations of dissolved CO_2 and H_2CO_3 in millimoles per litre to P_{CO_2}; S is temperature dependent.

PHYSIOLOGICAL REGULATION

The principal buffer system in the plasma and extracellular fluid compartments is the bicarbonate/carbonic acid system. The hydrogen ion concentration in those compartments is controlled by the mechanisms which regulate the concentrations of bicarbonate and carbonic acid (as P_{CO_2}) which are the renal and respiratory mechanisms respectively. In these two mechanisms:

The P_{CO_2} is maintained in a steady state by the balance struck between the rate of its metabolic production and the rate at which the lungs remove CO_2 from the body.

The HCO_3' concentration is regulated by the kidney through the activity of the epithelium in the proximal and distal renal tubules.

Respiratory Control

An increase in hydrogen ion concentration due to the metabolic production of CO_2, and consequent formation of carbonic acid, is dealt with by the respiratory control mechanism.

The arterial P_{CO_2} is maintained in a steady state by the balance struck between the rate of CO_2 production and the rate of its removal from the body by the lungs. The concentration of CO_2 in the alveolar spaces is inversely proportional to the rate of alveolar ventilation. The P_{CO_2} of arterial blood is in equilibrium with the CO_2 in the alveolar spaces and thus arterial P_{CO_2} is determined by the rate of alveolar ventilation. The rate of alveolar ventilation is governed by the respiratory centre acting under the influence of chemoreceptors and is located in the ventral surface of the medulla. The chemoreceptors are sensitive to changes in the hydrogen ion concentration of their bathing fluid; an increase or decrease in hydrogen ion concentration being respectively associated with an increase or decrease in ventilation rate. The magnitude of a change in hydrogen ion concentration following any given change in P_{CO_2} is determined by the buffering ability of the medium in which the change occurs. Cerebrospinal fluid is poorly buffered and an increase in P_{CO_2} is accompanied by a relatively large increase in hydrogen ion concentration. An increased P_{CO_2} in the arterial blood which perfuses the area of the respiratory chemoreceptors leads directly to an increase in the hydrogen ion concentration with a consequent stimulation of the respiratory centre. The P_{CO_2} of arterial blood is therefore both controlled by and is also the controlling mechanism involved in normal alveolar ventilation.

The respiratory centre also responds to changes in whole body hydrogen ion status associated with disease states. As increase or decrease in hydrogen ion concentration leads to an increase or decrease in the ventilation rate respectively. In the former situation with an increase in hydrogen ion concentration the consequent fall in bicarbonate concentration in the buffering reaction leads to a change in the HCO'_3/P_{CO_2} ratio; the increase in ventilation rate leads to a reduction in P_{CO_2} concentration with a reversion of the disturbance in the ratio and hence a *compensatory* correction of pH.

Owing to the respiratory control of P_{CO_2}, any change in either the rate of CO_2 exchange across the alveolar membrane or between the alveolar air and external air, which is directly attributable to primary pulmonary diseases, can of itself cause a disturbance in acid-base steady state (see Respiratory Acidosis and Alkalosis).

Renal Control

The strong acids that are formed either in the cellular metabolism of neutral dietary and tissue constituents, or are ingested in the diet as preformed or potential acids, dissociate into hydrogen ions and their weak conjugate bases. In a steady state net hydrogen ion production = renal net hydrogen ion excretion. A normal individual on a normal mixed diet has to excrete approx. 60 mmol of hydrogen ion per day to maintain steady state.

In the cells of the *proximal renal tubule* the process of the *"reabsorption" of filtered bicarbonate* takes place. This process does not represent a route of hydrogen ion excretion from the body, but is the mechanism whereby bicarbonate ions filtered at the glomerulus in the glomerular ultrafiltrate are "reabsorbed" by the tubules. This mechanism is responsible for the "reabsorption" of approx. 90% of the bicarbonate ions that have been filtered in the glomerular ultrafiltrate. The actual bicarbonate ions that have been filtered and enter the ultrafiltrate, and the lumen of the tubule, are not reabsorbed as such in this mechanism; the other feature of this process is that it does not involve any change in the pH of the glomerular ultrafiltrate. In this latter respect the pH of the fluid in the furthermost part of the lumen of the tubule is the same as that entering it from the glomerulus.

In this mechanism (Fig. 1) sodium and bicarbonate ions enter the lumen of the proximal renal tubule in the glomerular ultrafiltrate. The sodium ions diffuse into the tubule cells down a concentration gradient and are actively extruded into the peritubular fluid by a pump mechanism which maintains intracellular sodium at a low value. Hydrogen ions generated in the tubule cells by carbonic anhydrase, enter the lumen of the tubule in exchange for the sodium ions and combine with the filtered bicarbonate to form carbonic acid. The latter dissociates into water and carbon dioxide; the carbon dioxide diffuses into the tubule cell where it takes part in the carbonic anhydrase mediated generation of hydrogen and bicarbonate ions.

The renal control of hydrogen ion concentration in the extracellular fluid compartment by variations in urine pH involves both the simultaneous *regulation* of bicarbonate concentration as well as the *excretion* of hydrogen ions from the body. In the extracellular fluid compartment any increase in hydrogen ion concentration is buffered by the bicarbonate/carbonic acid system with a consequent reduction in bicarbonate concentration. The *distal renal tubule* mechanism not only excretes the excess of hydrogen ions but also simultaneously generates bicarbonate ions to replace those utilized in the initial buffering reaction. These two functions take place within the cells of the *distal renal tubule*. In these cells, under the influence of *carbonic anhydrase* (CA), hydrogen ions are generated by dissociation from carbonic acid:

$$CO_2 + H_2O \overset{CA}{\rightleftharpoons} H_2CO_3 \rightleftharpoons H^+ + HCO_3'$$

The hydrogen ions are secreted into the tubule lumen in exchange for Na^+ ions while the HCO_3' ions are secreted into the peritubular fluid.

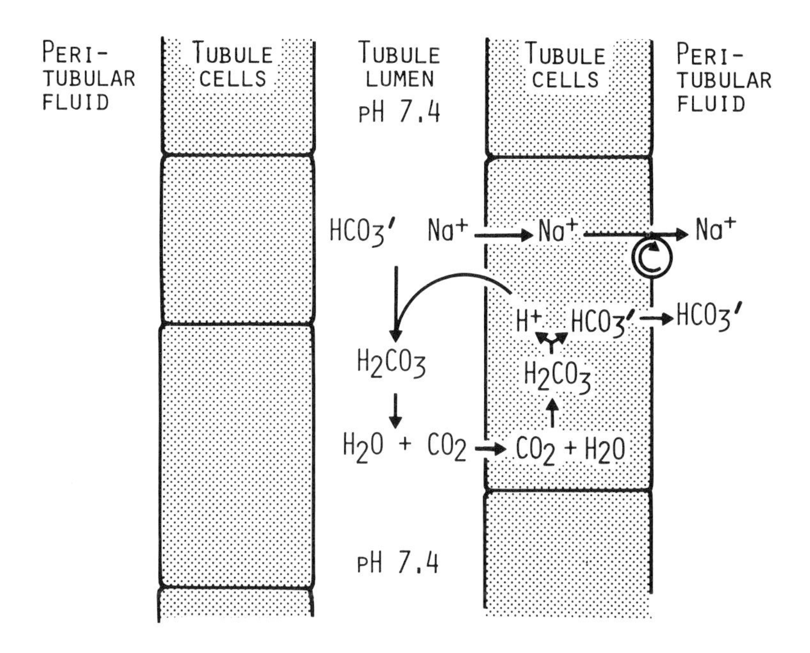

Fig. 1. *Mechanism of bicarbonate "reabsorption" by the cells of the proximal renal tubule.*

The distal renal tubules excrete hydrogen ions in the urine by three mechanisms (Fig. 2).

(1) *Excretion of free hydrogen ion.* This mechanism is very inefficient. The hydrogen ion concentration of a litre of urine with a pH of 4 is 0.1 mmol/litre. If, therefore, 2 litres of urine were produced each day with a pH value of 4 then only 0.2 mmol of hydrogen ion would be lost from the body as free hydrogen ion; a negligible amount in view of the total normal daily production rate of 60 mmol.

(2) *Excretion of ammonium.* Ammonia is generated in the cells of the renal tubules from glutamine and amino acids by glutaminase and oxidative deamination respectively. The ammonia diffuses out from the cells of the tubule into the lumen where it combines with hydrogen ion to form ammonium; the hydrogen ion having been simultaneously generated in cells of the tubules by the carbonic anhydrase system. The mechanism for the excretion of ammonium by the renal tubules simultaneously removes hydrogen ion from the body, replaces a bicarbonate ion and thus maintains steady state.

(3) *Excretion of titratable acid.* Hydrogen ions generated in the renal tubule cells, under the influence of carbonic anhydrase, are secreted into the lumen of the renal tubules and combine with conjugate bases in the tubular fluid converting them to their conjugate acids. Hydrogen ion secretion is linked with active sodium transport. The conjugate acids constitute the *titratable acid* of urine and the mechanism of the conbination of the conjugate bases (or buffers) with hydrogen ion is also termed the *titration of buffers.* The monobasic/dibasic phosphate buffer pair (H_2PO_4'/HPO_4'') plays the dominant role in the renal excretion of hydrogen ion by this mechanism. At the normal pH of the glomerular ultrafiltrate the ratio of dibasic to monobasic phosphate is 4:1. In the lumen of the tubules the secreted hydrogen ions convert the base from the dibasic to the monobasic form:

$$H^+ + HPO_4'' \rightleftharpoons H_2PO_4'$$

If sufficient hydrogen ions are secreted by the tubule cells the urine pH can be reduced to a value of 4.5, at which value almost all of the phosphate is in the monobasic form. The rate of excretion of hydrogen ions from the body by this mechanism is proportional to the rate of excretion of conjugate base in the urine. If the buffer

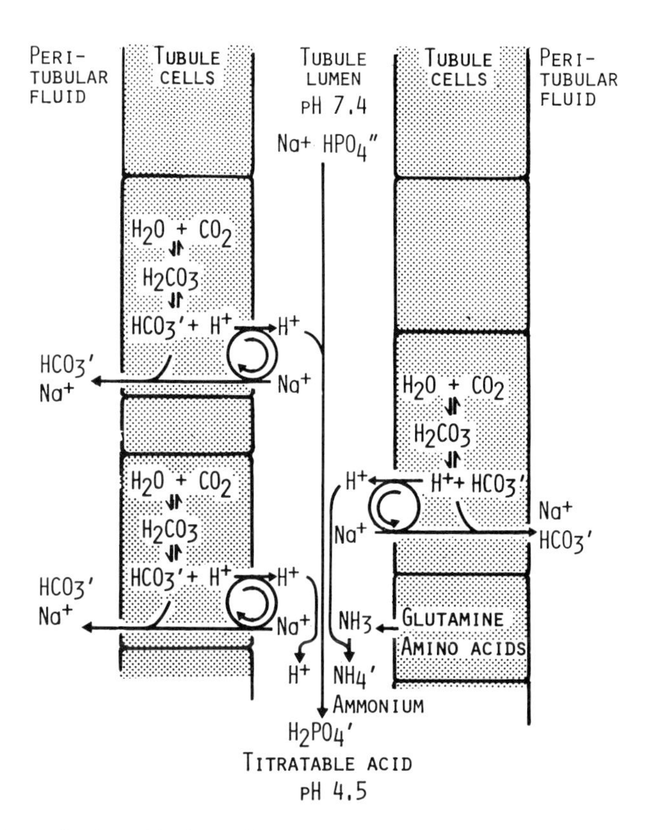

Fig. 2. *Mechanism of hydrogen ion excretion by the cells of the distal renal tubule. (Reproduced from Wills, M. R. (1978). "The Metabolic Consequences of Chronic Renal Failure" (2nd ed.) by kind permission of HM & M Publishers Ltd., Aylesbury, U.K.)*

capacity of the urine is high a large amout of hydrogen ion can be removed from the body by this mechanism. As in the preceding mechanism hydrogen ion secretion into the lumen of the tubules is associated with the simultaneous replacement of a bicarbonate ion in the extracellular fluid space.

CLINICAL DISORDERS

Clinical disorders which cause a disturbance in the regulation of hydrogen ion concentration in the extracellular fluid compartment lead to either acidosis or alkalosis.

Acidosis

Acidosis is strictly defined as an increased concentration of hydroxonium ions (H_3O^+) in the extracellular fluid compartment. The hydroxonium ion is the form in which hydrogen ion exists in an aqueous phase:

$$H^+ + H_2O \rightleftharpoons H_3O^+;$$

as such the increase in hydroxonium ions represents an increase in hydrogen ions. The usual *clinical definition of acidosis* is an excessive hydrogen ion concentration in the extracellular fluid compartment.

Clinical states of acidosis may be divided into *respiratory* where the "acid" source of the hydrogen ion is H_2CO_3, and *metabolic* where the "acid" source is from any acid other than H_2CO_3. *This sub-division is based on the respiratory and renal regulation of hydrogen ion status by CO_2 disposal and renal excretion.*

The *mechanism* involved in the aetiology of acidosis in all clinical situations must be explicable either as the retention of hydrogen ions or as the excessive loss of base, with a consequent relative increase in hydrogen ion concentration. A concept of the aetiology of acidosis in all clinical situations involves a consideration of:

sources of hydrogen ions;
mechanisms for the disposal of hydrogen ions;
derangements that disturb the normal excretion of hydrogen ions.

An overall physiological schema of the production, buffering and excretion of hydrogen is illustrated in Fig. 3. This schema although being considered here in the concept of acidosis is equally applicable to alkalosis.

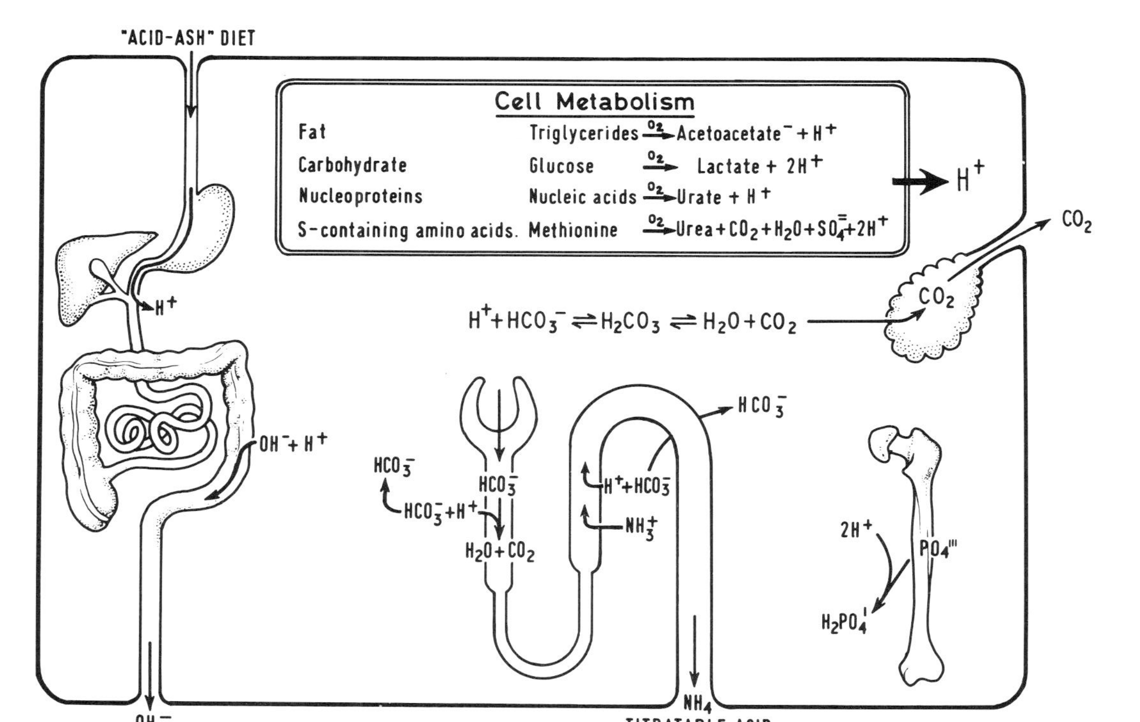

Fig. 3. *An overall schema for the sources of hydrogen ions, their buffering and the mechanisms for their disposal. (Reproduced from Wills, M. R. (1978). "The Metabolic Consequences of Chronic Renal Failure" (2nd ed.) by kind permission of HM & M Publishers Ltd., Aylesbury, U.K.)*

Sources of Hydrogen Ions

Carbon dioxide. Carbon dioxide produced in the tissues in cellular metabolism is normally transported in the blood to the lungs and excreted at a rate which equals that of production. Carbon dioxide exists in the extracellular fluid compartment in the dissolved form which can react with water to form carbonic acid and as such represents a potential source of hydrogen ions:

$$CO_2 + H_2O \rightleftharpoons H_2CO_3 \rightleftharpoons H^+ + HCO_3'.$$

Cell Metabolism of Neutral Dietary and Tissue Constituents.
Carbohydrates

$$\text{Glucose} \xrightarrow{O_2} 2\ \text{lactate} + 2H^+$$

Fats

$$\text{Triglycerides} \xrightarrow{O_2} \text{Acetoacetate} + H^+$$

Nucleoproteins

$$\text{Nucleic acids} \xrightarrow{O_2} \text{Urate} + H^+$$

Sulphur containing amino acids

$$\text{Methionine} \xrightarrow{O_2} \text{Urea} + CO_2 + H_2O + SO_4'' + 2H^+$$

Dietary intake of preformed or potenital acid. Compounds under this heading are those that constitute what is often termed the "Acid-Ash" nature of the diet: it includes phosphoproteins (e.g. phosphoserine), phospholipids (e.g. lecithin) and various metabolizable cations (e.g. arginine).

Mechanisms for the Disposal of Hydrogen Ions

The kidneys and lungs jointly participate in the regulation of hydrogen ion concentration in the extracellular fluid compartment by hydrogen ion excretion and by stabilization of the bicarbonate/carbonic acid buffer system. In this joint mechanism the kidneys are responsible for the excretion of hydrogen ions and the regeneration of bicarbonate ions while the lungs are responsible for the removal of CO_2. On the basis of this conceptual approach clinical states of acidosis and disturbances of hydrogen ion excretion are divided into *respiratory* and *metabolic.* In the respiratory group the primary defect is in the maintenance of arterial P_{CO_2} concentration, with a consequent increase in carbonic acid and hydrogen ion concentration. In the metabolic group the primary disturbance can be in the intake, production or excretion of hydrogen ions.

Respiratory acidosis. This may be defined as an increase in hydrogen concentration in the extracellular fluid compartment due to a failure to excrete carbon dioxide.

The *clinical states* in which respiratory acidosis may occur can be sub-divided into those which involve pulmonary gaseous exchange and those which involve central respiratory depression, as shown in Table 1.

Table 1 *Clinical conditions which are commonly associated with respiratory acidosis*

Impairment of gas exchange
(i) across the alveolar membrane
Cardiac failure with pulmonary congestion
Pneumonia
Infiltrating tumours
Emphysema
(ii) between alveolar air and the external air
Lesions involving the nerve supply to the respiratory muscles (poliomyelitis, myasthenia gravis etc.)
Impairment of respiratory muscle action (thoracic deformity, muscle damage etc.)
Obstructive airways disease; upper or lower (asthma, tracheal stenosis etc.)
Depression of respiratory centre
Morphine and other central depressant drugs
During anaesthesia
Cerebral injury
Cerebro-vascular accidents

The *clinical features* of respiratory acidosis are directly attributable to the effects on the central nervous system of the retention of carbon dioxide and the consequent increase in the hydrogen ion concentration of the cerebrospinal fluid. Cerebrospinal fluid is relatively poorly buffered and since CO_2 readily diffuses across the blood-brain barrier an increase in arterial P_{CO_2} is rapidly associated with marked changes in the pH of cerebrospinal fluid. The clinical features associated with respiratory acidosis are collectively described in the syndrome referred to as "*carbon dioxide narcosis*". In the early stages the syndrome is characterized by *cyanosis* with *hyperventilation at rest, fatigue* and *weakness,* with progression to *impairment of consciousness, twitching of fingers at rest,* and finally to *delirium* and *coma.*

Table 2 *Clinical conditions which are commonly associated with metabolic acidosis*

Excessive intake of hydrogen ions
- NH_4Cl poisoning (or ingestion of HCl, methionine or arginine monohydrochloride)

Excessive production of hydrogen ions
- Diabetic ketoacidosis
- Lactic acidosis
- Ketoacidosis or starvation

Defective elimination of hydrogen ions
- Renal failure; acute or chronic
- Disorders of the renal tubules with failure of the sodium for hydrogen ion exchange pump with a consequent *renal tubular acidosis* and secretion of an alkaline urine; associated with an excessive loss of Na^+, or Na^+ and K^+, the Lightwood and Albright types respectively.
- Treatment with diuretic drugs that act by inhibiting renal tubular carbonic anhydrase activity
- Adrenal insufficiency -- a failure of the sodium for hydrogen ion exchange pump mechanism

Excessive loss of base
- Diarrhoea
- Uretero-colic anastomosis
- Biliary and pancreatic fistulae
- Ileostomy

Cerebrospinal blood flow is increased with *raised intracranial pressure* and *papilloedema*, with clinical symptoms of *headache* and *blurring* of *vision*.

Metabolic acidosis. This may be defined as an increased hydrogen ion concentration in the extracellular fluid compartment due either to excessive intake, excessive production or defective elimination of hydrogen ions, or to an excessive loss of base with a consequent relative increase in hydrogen ion concentration.

The *clinical states* in which metabolic acidosis may occur may be sub-divided according to the differing aetiological mechanisms, as shown in Table 2.

The clinical features of metabolic acidosis are difficult to define precisely as in most, if not all, clinical situations it is difficult to separate the associated biochemical effects of the condition causing the metabolic acidosis (e.g. chronic renal failure, diabetes mellitus, etc.) from the pure effect of an increased hydrogen ion concentration on cell function. A *mild degree* of metabolic acidosis may be relatively symptomless or may be accompanied by *loss of appetite, nausea, headache and lethargy.*

Severe degrees of metabolic acidosis may be accompanied by a variety of clinical features, some of which are directly attributable to the increased hydrogen ion concentration.

Air hunger (Kussmaul breathing). An increased depth and frequency of respiration as a direct consequence of the effect of an increased hydrogen ion concentration on the respiratory centre. The depth of respiration may be increased prior to any change in the rate of respiration.

Lethargy, impairment of consciousness with progression to *coma.* The effects of an increased hydrogen ion concentration on the central nervous system are more closely linked with the pH value of the cerebrospinal fluid rather than with the pH of the arterial blood.

Peripheral vasodilatation. Initially the cardiac output may be increased but with severe acidosis it falls and *hypotension* may be pronounced.

Skeletal decalcification. If metabolic acidosis persists over a long period of time (e.g. chronic renal failure) a loss of bone mineral mass may occur which is due to the mobilization of phosphate base. This mechanism comes into operation as part of the whole-body buffering mechanism to counteract the hydrogen ion retention.

Alkalosis

Alkalosis is strictly defined as a decreased concentration of hydroxonium ions in the extracellular fluid compartment; however, the usual *clinical definition of alkalosis* is a decrease in hydrogen ion concentration in that compartment. The mechanism involved in the aetiology of alkalosis in all clinical situations must be explicable either as the excessive loss of hydrogen ions or as the retention of base, with a consequent relative decrease in hydrogen ion concentration. As with a concept of acidosis in clinical situations any consideration of the aetiology of alkalosis involves the mechanisms for both hydrogen ion production and disposal and these have already been discussed. The derangements that disturb the normal excretion or control of hydrogen ion concentration in the extracellular fluid compartment are divided, on the same basis as for acidosis, into respiratory and metabolic.

Respiratory alkalosis. This may be defined as a decrease in hydrogen ion concentration in the extracellular fluid compartment as the result

Table 3 *Clinical conditions which are commonly associated with respiratory alkalosis*

Increased gas exchange across the alveolar membrane, and between the alveolar air and external air
During anaesthesia
Artificial ventilation
Hysterical overbreathing or idiopathic hyperventilation
Stimulation of the respiratory centre
Fever
Cerebral injury
Drugs with central stimulating effect

of a reduction in arterial P_{CO_2} with a consequent fall in carbonic acid concentration.

Clinical situations in which respiratory alkalosis occurs may be sub-divided into the conditions shown in Table 3.

In all of these clinical situations there is hyperventilation with loss of CO_2 in the expired air, a reduction in arterial P_{CO_2} and a consequent fall in hydrogen ion concentration in the extracellular fluid compartment.

The *clinical features* of respiratory alkalosis are predominantly those of an *increased irritability of the central and peripheral nervous systems.* This is attributable to the role of hydrogen ion in maintaining the motor end-plate potential in nerve conduction and the loss of this effect in situations where there is a reduced hydrogen ion concentration. Since the primary mechanism in respiratory alkalosis is a reduction in arterial P_{CO_2} and since CO_2 readily diffuses across the blood-brain barrier, with a consequent rapid change in cerebrospinal fluid pH, the onset of tetanic symptoms in respiratory alkalosis may be more rapid and dramatic than for a similar degree of metabolic alkalosis as reflected by the change in the patient's arterial pH.

Metabolic alkalosis. This may be defined as a decrease in hydrogen ion concentration in the extracellular fluid compartment due either to excessive loss or to increased hydrogen ion excretion, or to an excessive intake of base. A metabolic alkalosis is particularly likely to occur during the treatment with absorbable antacids in the presence of renal failure (milk-alkali syndrome). In treatment with non-absorbable antacids such as aluminium hydroxide, sodium hydroxide is formed in the gastro-intestinal tract and there is consequently absorption of hydroxyl ions from the latter.

Table 4 *Clinical conditions which are commonly associated with metabolic alkalosis*

Excessive loss of hydrogen ions
Vomiting in pyloric stenosis
Gastric suction or fistula
Congenital chloridorrhoea
Increased hydrogen ion excretion
Diuretic drug therapy
Therapy with sodium retaining steroids
Cushing's syndrome
Primary aldosteronism
Potassium depletion – (in this situation there is an associated and consequent loss of hydrogen ions into the intracellular fluid compartment and into the urine as a direct result of the potassium deficiency)
Excessive intake of base
Therapy with absorbable and non-absorbable antacids (especially in the presence of renal insufficiency, e.g. milk-alkali syndrome)
Therapy with non-absorbable antacids (aluminium hydroxide, sodium hydroxide)

Clinical situations in which metabolic alkalosis occurs may be sub-divided according to the differing aetiologies shown in Table 4.

$$Al(OH)_3 + Na_2HPO_4 \rightleftharpoons Al_2(HPO_4)_3 + NaOH$$

The *clinical features* of metabolic alkalosis are attributable primarily to the fall in hydrogen ion concentration in the extracellular fluid compartment and also to the associated hypokalaemia. The predominant effect of the reduction in hydrogen ion concentration is an *increased irritability of the central and peripheral nervous systems.* The increased irritability of the nervous system is manifested by *numbness* and *parasthesiae* progressing to *mental confusion* and *tetany.* The tetany associated with metabolic alkalosis is indistinguishable from that occurring in association with hypocalcaemia. A reduction in hydrogen ion concentration causes a direct depression of the respiratory centre with *slow, shallow respirations.*

In hydrogen ion depletion there is an increased urine loss of potassium and also a shift of potassium from the extracellular to the intracellular fluid compartment. The clinical features of the *consequent hypokalaemia* are identical to those found in potassium deficiency from any other cause and include *weakness, paralytic ileus, tachycardia, various cardiac arrhythmias*, and increased *susceptibility to digitalis intoxication.*

Secondary or Compensatory Changes in Acidosis and Alkalosis

The secondary or compensatory changes, associated with either an increased or decreased hydrogen ion concentration in the extracellular fluid compartment, involve alterations in whole-body hydrogen ion status and are reflected by changes in the intracellular and extracellular buffer pair systems and the mobilization of bone salt.

In *respiratory disturbances* with acidosis the *renal tubules* undertake *compensation* by the increased excretion of hydrogen ions in the urine. The compensation is, however, *always incomplete* and the blood pH is low in "compensated" respiratory acidosis. Similarly in respiratory alkalosis renal tubular compensation takes place with a decreased reabsorption of bicarbonate and the excretion of an alkaline urine but, as before, only partial compensation takes place.

In *metabolic disturbances* alterations in *respiratory rate* may partially *compensate* for the change in extracellular fluid hydrogen ion concentration by either a reduction or an increase in CO_2 excretion. These changes are reflected as alterations in arterial P_{CO_2} which tend partially to restore towards normal the bicarbonate/carbonic acid buffer-pair ratio.

Bone salt presents an abundant reserve source of the monobasic/dibasic phosphate buffer-pair system. In acidosis of all types mobilization of bone salt may occur as part of the whole-body hydrogen ion homeostatic system. The control and role of bone salt mobilization in this situation is not clear and is also controversial; it may involve either a direct or an indirect cellular mediated mechanism.

Diagnosis of Clinical Disturbances

The clinical history and examination (to define routes of fluid loss, drug therapy, etc.) should help to indicate if the *primary* nature of an acid-base disturbance was either respiratory or metabolic. The biochemical diagnosis is based on the simultaneous measurement in arterial blood of the values of pH, bicarbonate and P_{CO_2}. It is also usually of value to measure PO_2 simultaneously. In primary respiratory disturbances the degree of respiratory rate change in relation to the blood bicarbonate value is usually greater than in the metabolic group of patients; in patients with severe respiratory acidosis the bicarbonate is unlikely to fall below 18 mmol/litre whereas in those with metabolic acidosis obvious hyperventilation does not occur until the bicarbonate value has fallen below 12 to 14 mmol/litre.

Nearly all clinical disturbances are *mixed*, in that there is the additional effect of the *secondary or compensatory phenomena*. In the interpretation of the blood biochemical findings in acidosis it is essential to adhere to the conceptual approach that the sources of the hydrogen ions in metabolic disturbances are from any acid other than H_2CO_3, while in respiratory disturbances the *primary event* is a failure of removal of CO_2 and hence the source of the excess of hydrogen ions is H_2CO_3. This approach, which is equally applicable to alkalosis, facilitates a *logical interpretation* of data and obviates the use of nomograms and similar unnecessary devices.

Interpretation of disturbances in clinical situations. In the consideration of the laboratory-measured pH, P_{CO_2} and bicarbonate data in clinical situations it is essential to consider the results; firstly, in conjunction with the *clinical history* and *findings on examination* and secondly, to remember that in metabolic disturbances the striking changes are in the *bicarbonate* value while in respiratory they are in the P_{CO_2} value and that *secondary compensation* is *never complete.*

Examples

(1) 62-year-old male with a long history of chronic bronchitis and emphysema. Admitted as an emergency with recent respiratory infection of three days' duration. On examination: cyanosed, difficulty in breathing and mentally confused.

Arterial blood results

pH	7.22	(7.38–7.42)
Pa_{CO_2}	82 mm Hg	(38–42)
Bicarbonate	34 mmol/litre	(24–32)

Interpretation: Respiratory acidosis (chronic pulmonary disease). Low blood pH marked CO_2 retention, consistent with a respiratory origin for the excess hydrogen ions, with an increase in bicarbonate due to secondary (renal) compensation.

(2) 55-year-old female, gravida 6, with a long history of recurrent urinary tract infections, now complaining of lethargy, anorexia, backache and shortage of breath on exertion. On examination: pale, deep "sighing" respiration with an increase in respiratory rate.

Arterial blood results

pH	7.16
Pa_{CO_2}	30 mm Hg
bicarbonate	12 mmol/litre

Interpretation: Metabolic acidosis (chronic renal failure). Marked hydrogen ion retention with low blood pH and reduction in bicarbonate, consistent with metabolic origin of the disturbance and a secondary (respiratory) compensation with a reduction in CO_2 content.

(3) 28-year-old female on the fourth post-operative day after thyroidectomy for goitre complaining of "tingling sensations" in fingers and around the mouth. On examination: increased irritability on tapping the facial nerve (+ve Chvostek's sign) and an increased respiratory rate; already shown to be normocalcaemic.

Arterial blood results

pH	7.58
Pa_{CO_2}	23 mm Hg
bicarbonate	20 mmol/litre

Interpretation: Respiratory alkalosis (idiopathic hyperventilation). An increased blood pH with a marked reduction in CO_2, consistent with a respiratory origin for the disturbance, and a secondary (renal) compensatory reduction in bicarbonate concentration.

(4) 58-year-old male with a long history of dyspepsia. Admitted as an emergency with a 10-day history of copious vomiting (5 to 6 litres a day) and abdominal pain. On examination: dehydrated, mentally confused and with some slowing of respiratory rate.

Arterial blood results

pH	7.60
Pa_{CO_2}	48 mm Hg
bicarbonate	52 mmol/litre

Interpretation: Metabolic alkalosis (vomiting with loss of hydrogen ions). An increased blood pH with marked increase in bicarbonate, consistent with metabolic origin of the disturbance, and increased P_{CO_2} content due to carbon dioxide retention as the result of secondary (respiratory) compensation.

Treatment of Clinical Disturbances

Respiratory Acidosis and Alkalosis

The treatment of either an excess or deficit of hydrogen ions in the extracellular fluid compartment associated with respiratory disorders is primarily that of the underlying disorder, although in some situations the treatment of the acid-base disturbance may be necessary.

Acidosis. In acute respiratory acidosis the retained CO_2 stimulates the respiratory centre with a resultant increase in both the rate and depth of respiration. If ventilation is efficient with adequate gas exchange, then this mechanism would of itself lead to a reduction in P_{CO_2}; in many situations, however, an acute onset ventilatory problem is the cause of the respiratory acidosis. The treatment of acute respiratory acidosis is therefore primarily the improvement of ventilatory efficiency and gas exchange. In patients with *chronic* CO_2 retention and high P_{CO_2} values the respiratory centre no longer responds to carbon dioxide and is "driven" by the arterial oxygen tension. The treatment of carbon dioxide narcosis with oxygen may remove that anoxic stimulus and prove fatal. It is essential in these patients to increase the elimination of carbon dioxide by improving ventilation and it is of value to simultaneously improve cerebral oxygenation with *cautious* oxygen therapy. The latter should, whenever possible, be undertaken only with guidance from an experienced chest physician. In those patients who are being given continuous oxygen therapy it is essential to monitor their therapeutic progress by repeated determinations of their arterial blood pH, P_{CO_2} and bicarbonate concentrations as well as their oxygen status (PO_2). If it is not possible to improve ventilation adequately and eliminate the excess of retained CO_2 then it is possible by the administration of carbonic anhydrase inhibitors to increase the renal excretion of bicarbonate and therapeutically create a hyperchloraemic metabolic acidosis; this renal manoeuvre is rarely necessary. The basis of this approach is to lower the circulating bicarbonate concentration and thus indirectly lower carbon dioxide content.

Alkalosis. If symptomatic, and causing tetanic or other phenomena of increased irritability of the peripheral nervous system, treatment aimed towards increasing blood carbon dioxide content is necessary; this can easily be accomplished in the conscious patient by a simple manoeuvre such as re-breathing into a paper bag.

Metabolic Acidosis and Alkalosis

In the treatment of metabolic disturbances of hydrogen status it is essential to institute simultaneously, whenever possible, the appropriate treatment for the causative mechanism.

Acidosis. The causative mechanisms which are amenable to therapy which simultaneously relieves the hydrogen ion disturbance are those where there is either an increase in hydrogen ion production or loss of actual or potential alkali. In these situations the primary therapy is therefore that of the causative mechanism. If the causative mechanism is not amenable to or is not likely to respond rapidly to specific therapy then the metabolic acidosis as such should be treated. A severe degree of metabolic acidosis is also of itself an indication for treatment. The treatment of metabolic acidosis involves the administration of alkali which is usually given as bicarbonate. The latter is given as the alkali of choice, usually as the sodium salt. The advantages of bicarbonate, over other compounds is that it directly replaces the lost bicarbonate ions. The disadvantage of bicarbonate is the hazard of inducing a hyperosmolar state, the consequence of the administration of an excess of sodium ions in the form of sodium bicarbonate. It is important therefore that bicarbonate requirements should be carefully calculated and therapy monitored. Alkali may also be given effectively as sodium lactate in many clinical situations, but its alkalinizing effect depends on oxidation of the lactate ion and it is of no value in patients with lactic acidosis. The administration of alkali should be withheld if there is the possibility that the patient may also have hypokalaemia; alkali will potentiate the cardiac effects of the latter and may induce a hypokalaemic cardiac arrest.

Calculation of bicarbonate requirements. The bicarbonate "space" in litres is equivalent to *approx.* 50% of total body weight (*not* water) in kilograms. In patients with very severe metabolic acidosis the apparent distribution "space" of bicarbonate may in fact represent a much larger percentage of body weight, and may even significantly exceed the body weight, although there is no evidence of either excessive hydrogen ion production or external bicarbonate loss. It has been suggested that in this situation the pre-existing plasma bicarbonate concentration in some way influences the apparent bicarbonate distribution "space". In such patients the administration of apparently adequate calculated amounts of bicarbonate may produce only small increments in the plasma concentration; regular monitoring is therefore essential during bicarbonate therapy. In the treatment of metabolic acidosis the calculation of bicarbonate requirements should be based on an amount which is sufficient to raise the plasma bicarbonate concentration to between 10 to 12 mmol/litre if the initial value is less than 10 mmol/litre or to between 16 to 18 mmol/litre in those patients with initial bicarbonate values between 10 to 12 mmol/litre. The total calculated amount of bicarbonate should be given by intravenous infusion over a minimum of 4 h; more rapid correction may cause tetany. This approach is based on the fact that in many patients their acid-base status is considerably improved by the simultaneous rehydration with volume replacement which improves cellular perfusion; any disturbance in cellular perfusion contributes to acidosis.

Example

50 kg patient with severe metabolic acidosis. pH 7.14, bicarbonate 11 mmol/litre: aim is to restore the plasma bicarbonate to approx. 16 mmol/litre, so the deficit is 5 mmol/litre.

bicarbonate "space" = approx. 25 litres*

total bicarbonate requirement = deficit × "space" = 5 × 25 = 125 mmol bicarbonate ions (which can be given intravenously as an isotonic solution)

A hazard of intravenous bicarbonate infusion is that a variety of drugs interact with this alkaline solution and may cause venous gangrene. In view of this hazard some authorities prefer to give intravenous bicarbonate slowly by syringe.

Specific Situations

The treatment of metabolic acidosis in patients with *diabetic ketoacidosis* depends upon the severity of the acidosis. In *mild* metabolic acidosis the administration of alkali is not essential as these patients usually respond to adequate rehydration and appropriate insulin dosage. In patients with a *severe* metabolic acidosis (pH less than 7.3) the administration of alkali may become essential as these patients may fail to respond to apparently adequate rehydration and insulin dosage in the absence of simultaneous treatment with alkali. The latter should

* Bicarbonate "space" is equivalent to approx. 50% body weight in kg.

not, however, be started until the patient's plasma potassium concentration is available. As well as being dehydrated many patients with diabetic ketoacidosis also have a whole body deficiency of both potassium and sodium. The initial or immediate intravenous fluid replacement therapy in these patients should consist of physiological saline, after an initial blood sample has been collected, but potassium and alkali should *not* be given until the initial pre-treatment plasma values are available. Alkali should not be given even if the patient is severely acidotic if the initial potassium concentration is 3 mmol/litre or less as this must be corrected first. Hypokalaemia may also develop with the institution of correction therapy for the primary carbohydrate abnormality and it is essential that as well as monitoring the blood glucose and ketone concentrations these patients must also have regular estimations of their arterial blood pH, P_{CO_2}, HCO_3' and electrolyte concentrations.

The treatment of metabolic acidosis in patients with *chronic renal failure* with intravenous alkali potentially presents the additional problems of volume overload and hypernatraemia. In such patients alkali therapy may be given orally.

The metabolic acidosis associated with *severe diarrhoea* is typically combined with hypokalaemia, as in this situation the alkali is lost as the potassium salt.

Alkalosis. As with all other disturbances of hydrogen ion status it is essential, if possible to institute treatment simultaneously for the causative mechanism. In the treatment of the metabolic alkalosis *per se* it is important to note that the intravenous infusion of physiological normal saline has an "*acidifying*" action, as also does potassium chloride. This effect is attributable to the interplay of sodium and chloride in the renal tubular handling of hydrogen ions.

In the renal tubule sodium is actively transported from the lumen across the cell into the peritubular capillaries. In this process electrical neutrality is maintained either by the transport of chloride as the accompanying anion or by the secretion of either hydrogen or potassium ions into the tubule lumen. Under normal circumstances approx. 115 mmol of sodium per litre of the glomerular ultrafiltrate are reabsorbed with chloride; the remaining 25 mmol are reabsorbed in exchange for either potassium or hydrogen ions. In the presence of hypochloraemia (for example in patients who are vomiting or who are on gastric suction), in order to maintain sodium status there is *enhancement* of the renal tubular sodium for hydrogen and potassium ion exchange pump, with the consequent development of hypokalaemic metabolic alkalosis. In the converse situation, as with a sodium chloride infusion, there is *suppression* of the exchange pump with a diminution in the renal tubular secretion of hydrogen ions and consequently the development of a metabolic acidosis; the "acidifying" effect of a saline infusion.

The most common cause of metabolic alkalosis is a reduction in total body potassium; long-term diuretic therapy without adequate potassium supplementation. In the majority of patients with metabolic alkalosis the infusion of sodium chloride, and adequate replacement of their potassium deficit is usually all that is required. It is important, however, to recognize that the correction of a metabolic alkalosis with the intravenous infusion of sodium chloride is slow and that *not* all clinical conditions are responsive. The latter are usually termed "*salt resistant*" and include hyperaldosteronism, Cushing's syndrome and severe potassium deficiency. In these conditions and in patients with severe metabolic alkalosis which warrants urgent treatment, if it is causing clinical symptoms, it is possible to administer hydrogen ions intravenously by a number of methods. These include ammonium chloride, arginine monohydrochloride or even hydrochloric acid. These techniques are *only rarely if ever needed.* The calculation of acid requirements in these situations is on the basis of the bicarbonate "space" and this has already been discussed.

Specific Situations

The metabolic alkalosis in patients who are either *vomiting* (as with pyloric stenosis) or are on *gastric suction* is usually associated with dehydration and hypokalaemia both of which require simultaneous correction therapy. The potassium depletion and hypokalaemia in this situation are not simply accounted for by loss in the gastric secretions (approx. 10 mmol/litre) but are the result of excessive urine loss, a direct consequence of the metabolic alkalosis. Potassium ions are secreted by the renal tubules in the sodium-exchange mechanism in lieu of hydrogen ions. In this renal exchange context it is of value to remember that "potassium deficiency causes a metabolic alkalosis and a metabolic alkalosis causes potassium deficiency". The primary factor which plays the major role in the genesis of the metabolic problems of patients who are either vomiting or are on continuous gastric suction is the excessive loss of hydrogen ions (approx. 60 to 80 mmol/litre of gastric juice). In these patients in the presence of *normal renal function* the immediate treatment should be rehydration with the intravenous administration of potassium, as chloride, according to the calculated requirements (on the basis of the whole body deficit) of these ions. In the presence of a severe metabolic alkalosis, or with *impaired renal function* or with an alkalosis that is unresponsive to the administration of sodium chloride, the administration of hydrogen ions may be necessary. In such patients with a severe refractory metabolic alkalosis who have a consequent respiratory depression arginine monohydrochloride is probably the method of choice.

Three

Nutrition and Metabolism

Gillian C. Hanson
Whipps Cross Hospital,
Leytonstone, London, UK

METABOLIC ASPECTS OF NUTRITION

For efficient growth, function, repair and maintenance the human body needs an adequate supply of water, carbohydrate, fat, protein, electrolytes and vitamins. Carbohydrate, fat (and protein) provide the energy needed for resting metabolism, synthesis of body tissues, excretory processes, maintenance of thermal balance, physical activity and specific dynamic action; while protein, vitamins, electrolytes and water play a structural role. Structure and function are mutually dependent and both rely on an adequate supply of energy and raw materials. It is useful to approach the subject of nutrition from this standpoint, bearing in mind also that the body is in a continual state of turnover and that no part of it can be regarded as a static entity.

Control of Metabolism

In metabolic terms the organism may be seen as a dynamic biochemical system in which there exists an equilibrium between catabolism and anabolism. This equilibrium is under hormonal control, ACTH, adrenaline, cortisol, glucagon and prolactin exerting a catabolic effect, while insulin promotes anabolism. The rate of hormone secretion depends in its turn on the nutritional state and metabolic demands at any given moment.

Basis Requirements

In order to remain in metabolic balance (i.e. catabolism = anabolism) the average healthy adult requires 30 kcal/kg/day (0.126 MJ/kg/day). It has been suggested that the total daily calorie requirements shall be provided as 70% from carbohydrate, 20% from fats and 10% from protein. Thus in a 79 kg man, the average requirements would be:

carbohydrate	350–370 g
fat	40–50 g
protein	45–55 g (7.2–8.8 g nitrogen)

These figures should not be regarded as absolute; the human body is extremely versatile in its needs, and provided certain minimum requirements are met it can adapt to wide variations in the relative proportions of its nutrients.

Metabolic Interrelationships

The three basic foodstuffs are carbohydrate, protein and fat. *Carbohydrates* are broken down to monosaccharides which may be utilized for energy, converted to the storage carbohydrate glycogen or, when consumed in excess, transformed into fat and stored as adipose tissue. *Proteins* are split up into their constituent amino acids. The latter may enter the body's pool of circulating amino acids from which they are immediately available for new protein synthesis or they may be degraded via gluconeogenesis to provide a source of energy, the nitrogeneous radicals being excreted in the urine. Ingested *fat* is broken down to free fatty acids and glycerol. These may be converted into ketone bodies to provide energy or, together with triglyceride derived from excess carbohydrate intake, may be laid down in the body's adipose store. It is important to remember that all these pathways are reversible and that there is continual turnover in all the compartments and stores mentioned. All the biochemical reactions concerned are in a constant state of flux, the overall direction of flow depending on the hormone balance.

Metabolic Response to Stress

It is important to recognize that the body reacts in different ways to various forms of metabolic stress.

In simple starvation, or primary input failure, without any gross underlying pathology plasma values of ACTH, adrenaline, cortisol, gluccagon and prolactin increase, causing the metabolic balance to "swing" in favour of the mobilization of lipid and carbohydrate from their respective storage forms, depot fat and glycogen. The absolute value of circulating insulin may also rise, but because of the simultaneous increase in catabolic hormones which antagonize the action of insulin its effect is decreased. The consequent stress of starvation is initially protein-sparing so that gluconeogenesis is avoided and nitrogen balance maintained for as long as possible.

In conditions such as major trauma or surgery the insulin restraint over protein breakdown is abolished and the swing to catabolism unchecked. In these situations the glycogen stores are rapidly depleted and there is a mobilization of fat and protein on a massive scale in an attempt to meet metabolic demands. Adrenal stimulation which occurs via the sympathetic nervous system in response to severe stress or trauma plays an important role in pushing the hormonal balance towards the extreme of unopposed catabolism.

Thus there are significant differences between the catabolic processes associated with simple starvation and those prompted by trauma and other pathological conditions.

Body Stores

Before considering specific clinical situations it is essential to have a concept of the nature, size and duration of the body's emergency reserves.

Glycogen is the only form in which the body can store carbohydrate. The major site of storage is the liver, although a small quantity is contained in muscle. Glycogen constitutes the sole source of rapidly available energy; the reserve is strictly limited in size, amounting to a total of only 300 to 400 calories (1.25–1.7 MJ). In acute situations this relatively small store can be depleted within 6 to 8 h after which, since carbohydrate is essential to maintain cellular metabolism, gluconeogenesis ensues resulting in rapid and excessive nitrogen losses.

The *adipose organ* is distributed widely throughout the body but mainly in the subcutaneous tissues and represents the major energy reserve. In a non-obese subject the average total body store is 12 to 15 kg and can provide energy for 20 to 25 days of simple starvation. The body is able to utilize ketone bodies as its primary fuel source, reserving glucose derived from gluconeogenesis for those tissues which are unable to adapt. After major trauma or surgery, however, where the catabolic process is augmented by the endocrine response, this protein-sparing mechanism is abolished. Thus, even in extreme obesity the patient's excessive lipid stores cannot protect him against overwhelming gluconeogenesis and nitrogen deficit in the face of serious trauma or major surgery; the need for nutrition is as great as that of his leaner counterpart. Nutritional supplementation should never be witheld because a patient needs to lose weight. The indications for supplementation are the same for fat and thin alike, and an acute life-threatening situation is not the ideal setting for instituting a weight-losing regimen.

The body's reserve store of free amino acids is small and in the absence of protein intake is utilized within a few hours; thereafter tissue protein must be broken down to maintain the amino acid pool. Protein plays a role as a structural element in tissue repair and

Table 1 *Approximate calorie and nitrogen losses in various clinical situations*

	Calories			
	MJ	kcal	Nitrogen g	(Protein g)
Chronic starvation	4.2–6.3	1000–1500	7.2–12.0	(45–75)
Post-operative major surgery without complications	8.4–12.6	2000–3000	12.0–20.0	(75–125)
Multiple trauma, or surgery with sepsis	12.6–18.9	3000–4500	20.0–32.0	(125–200)
Major burns	14.5–25.0	3500–6000	32.0–56.0	(200–350)

maintenance; but from the nutritional standpoint its major function is that of an emergency fuel reserve. Most of the protein available for this purpose is located in the muscle mass. This reserve will last for 10 to 12 days but its utilization results in generalized wasting, impairment of healing and tissue repair and nitrogen deficit. Thus, although protein is essential for tissue repair and enzyme formation, in catabolic situations where the organism is under severe stress the hormone balance changes, with the result that amino acids are diverted away from protein synthesis to be used as a primary energy source. In this situation tissue protein is catabolized to glucose and the nitrogenous radicals are excreted in the urine, 300 g of muscle tissue being lost for every 10 g of urinary nitrogen (1 g nitrogen = 6.26 g protein).

Nitrogen Losses and Nutritional Requirements

In a situation of simple input failure without additional metabolic demands daily nitrogen losses are approx. 7.2–12 g (45–75 g protein) while the energy deficit amounts to 4.2–6.3 MJ (1000–1500 kcal). The approximate calorie and nitrogen losses in various clinical situations are shown in Table 1.

It must be recalled that there is a 10–12% energy increase for every 1°C rise above normal body temperature. After minor injury or simple elective surgery the metabolic response tends to diminish after 4–8 days, but after major trauma, complicated surgery, or in severe sepsis; the response may continue unabated for many days and may indeed accelerate.

In the assessment of a patient's nutritional needs it is essential to take into account not only the nature of the pathology but also the nutritional status of the patient pre-operatively or before the acute situation arose.

It would be a mistake to assume that all patients require nutritional supplementation. Indeed most patients coming to elective operation can survive for a few days without nutrition and incur no serious consequences. It is important to be aware of the patients nutritional state and to recognize the indications for nutrition. Nutrition should be commenced early rather than late; it is easier to maintain nutrition than to replace lean body mass that has been lost.

Rationale for Nutrition During Acute Illness

A patient who loses more than one-third of their total body protein during an acute illness is likely to die from malnutrition alone. Protein losses are such that this may occur within 10 days in a patient suffering from severe burns and within 14 days in a post-operative patient who is suffering from serious complications such as sepsis. This fact has to be kept constantly in mind whenever treating a critically ill patient.

Clinical research has shown that a critically ill patient maintained in a good metabolic and nutritional state – is less susceptible to infection, is less likely to experience skin breakdown, the rate of mobilization is increased and the period of recovery is diminished.

FEEDING VIA THE GASTRO-INTESTINAL TRACT

The best method of feeding a patient who is unable to take food by mouth is via a Ryles tube. On rare occasions feeding may be introduced via a gastrostomy or duodenostomy. This booklet is not the place to discuss oral nutrition in detail and therefore only a few brief comments will be made.

(1) Elemental diets are very expensive and are generally unnecessary – a Ryles tube diet can be made up of simple constituents, namely Albumaid® (a beef hydrolysate), Caloreen® (a glucose polymer and therefore producing a lower osmotic effect) and occasionally the addition of fats such as neutral triglycerides.

(2) A patient who is unwell will not absorb a severely hypertonic feed and therefore the initial feeds must be well diluted. A hypertonic feed may lead to severe diarrhoea with consequent dehydration and potassium loss.

(3) Feeding must be started gradually, and the full diet should be built up over a period of days. Caloreen is generally the substance which will be tolerated best at first – fats should not be introduced until at least 5 days have lapsed from starting Ryles tube feeding.
(4) Vitamin supplements and haematinic supplements can generally be crushed or given in liquid form. Iron should be added at about day 5 since it may produce gastric irritation or induce vomiting. Vitamin B_{12} must be given intramuscularly and if there is any evidence of a malabsorption state, fat soluble vitamins should be given parenterally.
(5) Some patients will not tolerate feeds given at hourly intervals, in such cases a continuous drip feed may be successful.
(6) In certain patients an adequate dietary intake is not possible via the gastro-intestinal tract and supplemental intravenous nutrition and electrolyte supplementation may be necessary. In such cases 1 litre of water containing 100–200 g Caloreen given daily may be tolerated via the Ryles tube, additional carbohydrate, fat, protein, vitamins and electrolytes being given intravenously.

INTRAVENOUS NUTRITION

Indications

Intravenous feeding should be considered if:
(1) There has been more than 10% loss of body weight within 6 weeks prior to hospital admission. These cases are generally suffering from starvation due to gastro-intestinal disease. Should a planned operation be considered, these patients should be fed for 1–2 weeks pre-operatively.
(2) The patient is moderately to severely catabolic (protein loss more than 12.8 g nitrogen (80 g protein)) and is unlikely to absorb adequate quantities of fluid and calories within 3–5 days of the onset of his illness.
(3) The patient suffers from a high bowel fistula without distal obstruction. Here, by stopping oral fluids and maintaining metabolic and nutritional balance intravenously, the volume of fistula loss decreases and healing may occur.
(4) Occasionally, as a result of bowel infarction, or inflammation, the loss of small bowel is so extensive that adequate absorption is not

possible by the oral route until adaptation has taken place, or has been so extensive that long-term intravenous feeding may have to be considered in order to maintain life. In the latter case, consultation with an expert experienced in long-term parenteral nutrition is essential since these patients present a very rare and unique problem.

(5) Rarer and rather controversial indications for intravenous feeding include: patients suffering from chronic malnutrition as a result of carcinomatosis, patients anorexic before and during non-surgical treatment for cancer, and patients refusing to accept feeding via a Ryles tube, e.g. anorexia nervosa.

Methodology

The technique of intravenous feeding requires skill and constant observation for any abnormal trend. Each patient requires individual care, access to more senior advice is essential – it is safer not to feed a patient parenterally than to do it badly.

Having decided to feed your patient intravenously, certain observations must be made before commencing nutrition:

(1) All electrolyte deficiences must be met and the daily requirements for the basic electrolytes (Na^+.K^+) must be known.
(2) Feeding should not be started until the patient has a normal calcium magnesium and inorganic phosphate level. Should these be abnormal, further advice must be sought.
(3) A baseline serum albumin is valuable. In acute starvation, should the serum albumin be less than 25 g/litre and the patient be acutely hypercatabolic, replacement with salt-free albumin or purified protein fraction should be considered.
(4) Serum B_{12} and folate levels should be taken prior to commencing therapy.
(5) A random blood glucose should be taken since if the level is greater than 8 mmol/litre, insulin may be required from the onset of intravenous glucose feeding (see p. 67).
(6) The acid-base state must be known – an alkalosis may indicate total body potassium depletion (which may exist in the presence of a low normal serum potassium). A mild acidosis should be observed carefully since it may be aggravated by intravenous feeding.

(7) Cardiac and renal function should be assessed to ensure that the patient can tolerate a normal fluid volume, potassium, sodium and protein load.

Intravenous Feeding Line

The intravenous catheter for feeding purposes should be put in by an expert using a fully sterile (and if possible non-touch) technique. The line is generally fed into the superior vena cava or into the right atrium. The site of insertion must be cleaned thoroughly with povidone iodine and sealed with a clear plastic spray. The infusion site should be taken down thrice weekly, or when there is any temperature, and should be preferably changed once a week. Any pyrexia with no definite source of infection is an indication to take blood cultures and to remove the catheter and culture the catheter tip.

One of the greatest dangers of intravenous feeding is a septicaemia related to catheter sepsis.

Investigations and Observations (Table 2)

The frequency of these observations depends upon the state of the patient – the greater the degree of catabolism, the more frequently the investigations are required.

Once intravenous feeding has been established, and the patient is fit enough to be on a ward, investigations may be done less frequently.

Comments about Investigations and Observations

Blood glucose. Glucose intolerance invariably occurs during stress, the blood glucose should be maintained between 5–10 mmol/litre. Should a dextrose/insulin regime be required, glucose monitoring using the Reflomat system* (preferable since it is more reliable in the higher range than the Ames reflectance meter) is essential.

Serium sodium. A low serum sodium may *not* reflect a fall in total body sodium (see p. 21).

Serum potassium. The serum potassium must be above 3.6 mmol/litre before commencing intravenous feeding, a fall to below 3.5 mmol/litre is an indication to stop feeding until the serum potassium has been restored to normal.

* Labora Mannheim.

Table 2 *Routine investigations and observations during the initial and subsequent management of a patient on intravenous nutrition*

Investigation	Frequency
Blood glucose	Initially before and after a glucose challenge (see text)
Serum sodium	Initially 1–2 X daily, subsequently daily
Serum potassium	Initially 2 X daily, subsequently daily
Serum magnesium	3 X weekly (unless level low at the onset)
Serum calcium	2 X weekly (unless level low at the onset)
Serum inorganic phosphate	3 X weekly (unless level low at the onset)
Acid-base statue	Initially 1–2 X daily, subsequently 2 X weekly
Nitrogen balance studies (Lee and Hartley, 1975)	Daily (see text), subsequently 3 X weekly
Routine observations	
Daily weighing if this is possible	
Fluid balance chart	
Investigations which may assist in metabolic and fluid control	
Right atrial pressure monitoring	
Urinary electrolytes	
Electrolyte analysis of any fistula loss	

Serum magnesium, calcium and inorganic phosphate. Serum magnesium and inorganic phosphate may fall to dangerously low levels during intravenous feeding. The serum magnesium may be low prior to starting nutrition because of fistula losses. Serum magnesium and inorganic phosphate commonly fall during chronic starvation. Serum calcium is rarely low unless there has been a severe acute loss (e.g. pancreatitis), or a chronic loss of calcium or depletion of vitamin D over many months.

Daily Weighing

Crude changes in body weight (> 1 kg daily) are related to changes in fluid balance. A gradual fall over a period of days is consistent with loss of muscle mass.

Right Atrial Pressure Monitoring (R.A.P. monitoring)

Measurement of the pressure in the right atrium gives an excellent guide to the dynamic blood volume. This technique should be used in the seriously ill where tolerance of fluid volume may be impaired or where there may be oliguria. The right atrial pressure may be raised in

Table 3 *Suggested plans for total parenteral nutrition*

Bottles				Contents/500 ml Bottle	MJ	(kcal)	Carbo-hydrate (g)	Nitrogen (g)	Fat (g)	Elements (mmol)						
										Na	K	Mg	Ca	Zn	Mn	PO_4
2	3	3	2	Vamin glucose[a]	1.35	(325)	50	4.7		25	10	0.8	1.3			
2				Synthamin 14[b]	0.97	(180)		7.15		36.5	30	2.5				15
1	1	1	1	Electrolyte solution A with 20% w/v dextrose[b]	1.7	(400)	100					14	13	0.04	0.02	
	1	1	1	Electrolyte solution B with 20% w/v dextrose[b]	1.7	(400)	100				30					30
1	1		1	Dextrose 50%	4.2	(1000)	250									
1		1	1	Intralipid 20%[a]	4.2	(1000)			100							
				Low sodium. Low protein	14.4	(3450)	550	9.4	100	50	50	15.5	16	0.04	0.02	30
				Standard	11.6	(2775)	350	11.1	100	75	60	16.4	17	0.04	0.02	30
				Fat-free	11.6	(2775)	600	14.1		75	60	16.4	17	0.04	0.02	30
				Increased calories. High protein	14.3	(3410)	450	23.7	100	123	80	20.6	16	0.04	0.02	30

[a] KabiVitrum Ltd. [b] Travenol Ltd.

patients with an increased pulmonary artery pressure, or right ventricular pressure, and where this is suspected, advice should be sought as to what is considered the safe range of right atrial pressure for that patient. The normal range for the right atrial pressure measured in the mid-axillary line, the patient lying horizontally, is 4–8 cm H_2O.

Choice of Solutions for Intravenous Feeding

A range of intravenous solutions suitable for intravenous feeding is shown in Table 3.

Carbohydrates

Glucose is the carbohydrate of choice – fructose and sorbitol may produce a lactic acidosis and sorbitol may produce a water diuresis. Glucose should be infused at a rate of not greater than 0.5 g/kg body weight/h. It is usual in an adult to infuse 40–50 g of glucose over a period of 2 h having taken a blood glucose before and 30 min after the infusion. If the blood glucose is $>$ 9.0 mmol/litre before starting the infusion, insulin will be required and if it rises to 10.0 mmol/litre after the infusion, insulin should be used. Should glucose intolerance be present and insulin be required, further advice must be taken. If facilities are not available for adequate glucose monitoring, patient transfer to an area where these facilities are readily available should be considered.

Should insulin be required, pork insulin in a dose of 1 unit per g of glucose is added to the concentrated glucose solution (20, 40 or 50%) in a 100 ml buretrol burette. The dosage of insulin may be steadily increased until the blood glucose is kept at a steady level of 5.0–10.0 mmol/litre during the glucose infusion. During a dextrose/insulin regimen, glucose monitoring is essential (generally 2 hourly), a blood glucose should also be taken 1 h after stopping the dextrose/insulin regimen because of the danger of hypoglycaemia. The dose of glucose given is generally 100 g for the first day, gradually increasing to a total of 200–550 g daily. In states of extreme catabolism where a high carbohydrate intake is necessary in order to provide sufficient calories per gram of nitrogen infused (fats being poorly tolerated), it is advisable to give carbohydrate intravenously continuously via the central line, the protein containing solution (if not glucose containing) being given with the carbohydrate via a Y connector. Should fats be tolerated, these may be given via a peripheral line.

Protein

The majority of protein containing solutions now available consist of pure crystalline amino acids. There are numerous pure crystalline amino acid preparations, the one of choice must be in the laevorotatory form, must contain the full spectrum of essential and semi-essential amino acids and must not contain an excess of the "stuffer" amino acids such as glycine. The preparation of choice is at present probably the glucose containing preparation, Vamin glucose. Where nitrogen losses are greater than 13 g daily, a preparation with a higher protein concentration should be considered, such as Synthamin 14. It must be remembered that this preparation contains phosphate ions.

The nitrogen (protein) requirements can be assessed according to Table 1 and can be confirmed with the use of the nitrogen balance study (see Table 4). An attempt should be made to give nitrogen (1 g nitrogen = 6.25 g protein) in the same amount as the calculated loss.

It must not be forgotten that Vamin glucose contains glucose and therefore insulin may be required during infusion of this fluid. Insulin is generally not added to the bottle but given as a bolus intravenously hourly.

At least 150 calories should be provided as fat or carbohydrate per g of nitrogen infused. In the catabolic patient where fat intolerance is common this may necessitate the infusion of high concentrations of glucose. The management of the hypercatabolic patient requires considerable experience and senior advice must be sought.

Fats

Fats are generally given intraveneously as the soya bean oil preparation, Intralipid. Fat tolerance is generally poor initially and fat is rarely started until at least 24 h after commencing intravenous nutrition. The starting dose is generally 0.5 g/kg body weight/24 h gradually increasing to a total of 2.0 g/kg body weight/24 h.

Fat must not be infused when the serum is lipaemic and blood for biochemistry and haematology should be taken when the serum is fat free. Fats should be avoided during the initial intravenous nutrition of the hypercatabolic patient. It is advisable not to use fats when there is a haemorrhagic diathesis.

Table 4 *Clinical assessment of nitrogen losses. (This method of calculation is unsuitable in the presence of severe diarrhoea or fistula losses when direct estimation of the nitrogen losses in these fluids is essential)*

Protein catabolism	= urea excretion (g/24 h) $\times \frac{6}{5} \times \frac{28}{60}$ = urea excretion (g/24 h) $\times$ 0.56 = A	For untreated patients and those receiving crystalline amino acids
Blood urea correction	= change in blood urea[a] (g/litre) $\times$ total body water (60% body weight in kg) $\times \frac{28}{60}$ = change in blood urea (g/litre) $\times$ body weight (kg) $\times$ 0.28 = body weight	
Urinary nitrogen correction (usually negligible)	= g/24 h = C	
Total nitrogen loss per 24 h	= A + B + C	

[a] Applies only to rise in blood urea.

Table 5 *Electrolyte preparations that may be required during intravenous feeding in adults (when certain ions provided in Table 3 prove to be insufficient)*

Electrolyte additive	Dose and frequency	Route
Magnesium sulphate 50% 2 mmol Mg^{++}/ml	According to serum magnesium. Maintenance requirement approx. 0.1–0.25 mmol/kg body weight daily	Intramuscular
Calcium gluconate 10% 0.25 mmol Ca^{++}/ml or calcium chloride 10% 0.45 mmol Ca^{++}/ml	According to serum calcium. The quantity supplied in Table 3 should be adequate for maintenance	Intravenous in carbohydrate solution
Sodium bicarbonate 8.4% 1 mmol HCO_3^-/ml 1 mmol Na^+/ml	According to acid-base balance, urine and plasma sodium levels	Intravenous
Dipotassium hydrogen phosphate (K_2HPO_4) 2 mmol K^+/ml 1 mmol HPO_4^-/ml	According to serum phosphate. The quantity supplied in Table 3 should be adequate for maintenance	Intravenous in carbohydrate solution
Potassium chloride 15% 2 mmol K^+/ml	According to serum potassium. According to acid-base status and potassium losses	Intravenous in carbohydrate solution

Metabolic and Fluid Control

During intravenous feeding, electrolytes, acid-base and fluid control is essential – the requirements for magnesium, phosphate and potassium are generally very much higher than anticipated. Magnesium and inorganic phosphate can generally be adequately replaced according to Table 3, the average requirements for phosphate being 20–30 mmol daily and that for magnesium 10–15 mmol daily. Potassium requirements are very variable but in the absence of heavy losses and where 200–300 g of glucose are being infused per 24 h, generally 120–160 mmol are required daily – this is preferably given in the glucose infusion.

The electrolyte additives which may be required are shown in Table 5. Should extra ions be required other than potassium and above those recommended in Table 3, further advice must be taken.

Haematinics and Vitamins (see Table 6)

Folic acid replacement should be started early. Vitamin and haematinic replacements should be started within one week of commencing

Table 6 *Vitamin and haematinic replacement during intravenous nutrition*

Solivito[a] 1 vial	
Thiamin mononitrate	1.24 mg
Sodium riboflavine phosphate	2.47 mg
Nicotinamide	10.0 mg
Pyridoxine hydrochloride	2.43 mg
Sodium pantothenate	11.0 mg
Sodium ascorbate	34.0 mg
Biotin	0.3 mg
Folic acid	0.2 mg
Cyanocabalamin	2.0 μg
Vitlipid adult[a] 1 ml	
Retinol palmitate corresponding to retinol	75.0 μg (250 iu)
Calciferol	0.3 μg (12 iu)
Phytomenadione	15.0 μg
Fractionate soybean oil	100.0 mg

[a] KabiVitrum Ltd.

intravenous nutrition (Table 6). The most convenient fat soluble preparation being Vitlipid* (which is added to the Intralipid*) and the water soluble preparation Solivito*. Solivito should be given as a bolus injection daily. Since the quantity of folic acid and vitamin B_{12} in this preparation may be inadequate, it is usual to give folic acid 6 mg intramuscularly daily and vitamin B_{12} 1000 μg intramuscularly fortnightly.

Design of an Intravenous Diet

Certain criteria have to be met when designing an intravenous regimen. These are summarized in Table 7. An approximate estimate of the nitrogen (protein) requirement for different catabolic states is shown in Table 3. Having assessed the nitrogen (protein) requirement, the total number of calories and the grams of glucose needed daily can be estimated. The most suitable dietary regime for an individual patient can generally be selected from Table 3. It is important before starting intravenous feeding that preliminary criteria are met. Once nitrogen balance studies are available, the calorific requirements and intravenous regimen can be modified accordingly.

The diet should be started gradually, so that any intolerance or

* KabiVitrum Ltd.

Table 7 *Guidelines for establishing an intravenous feeding regimen in an adult*

Provide nitrogen according to Table 3
Provide 0.5–0.84 MJ (150–200 kcal)/g of nitrogen infused[a]
Provide 30–60% of calories as carbohydrate[a]
Intralipid 1–4 g/kg body weight/24 h. Ensure fat tolerance
Provide approx. 5 mmol K^+/g nitrogen infused
Ensure all electrolyte requirements are met
Ensure fluid balance and acid-base balance is correct
Supply vitamins and haematinics
Reassess requirements daily
(1 gN = 6.25 g protein)

[a] In the patient losing more than 12 gN a day, at least 200 cal/g of nitrogen infused should be provided, 50% of this being as carbohydrate.

metabolic abnormality is detected early. A full intravenous feeding regimen is rarely achieved in less than three days from commencement. Glucose tolerance must be assessed continuously, especially if the patient is on a dextrose/insulin regime since, as the patient becomes less catabolic, insulin requirements decrease. Any abnormal metabolic trend is an indication to stop intravenous feeding and to seek further advice.

Complications

Complications which may arise during intravenous feeding are enumerated below:

- Hyperosmolar coma
- Hypokalaemia
- Hypomagnesaemia
- Hypophosphataemia
- Folic acid deficiency
- Essential fatty acid deficiency
- Catheter sepsis leading to septicaemia

Hyperosmolar Coma

This is a condition where the serum osmolality is allowed to rise above normal and if allowed to rise above 310 mosmol/kg body water, may lead to severe cerebral dehydration and death from cerebral haemorrhage. Most intravenous feeding solutions are hypertonic and must therefore be given slowly in order to allow for metabolism. Monitoring of blood glucose levels and the appropriate use of insulin will prevent the development of hyperglycaemia. A rising serum sodium is a condition which must be recognized early and may be due to the inability of the kidney to clear sodium ions, or more commonly, to the excessive use of sodium containing solutions when the patient is suffering from predominant water loss (e.g. diarrhoea). It must be remembered that antibiotics may be given as the sodium salt, thereby increasing a sodium load, and that steroids are sodium retaining.

Hypokalaemia, Hypomagnesaemia and Hypophosphataemia

These are well-known complications of high glucose feeding. Hypomagnesaemia and hypophosphataemia should not arise if the regimens advised in Table 3 are adopted. Phosphate depletion is particularly likely to arise during high glucose feeding (400–600 g daily).

Folic Acid Depletion

This is most frequently seen when the patient is receiving drugs which are folate antagonists, is receiving intravenous ethanol as a nutritive source, or is on renal dialysis. The initial manifestations are generally thrombocytopenia and leucopenia.

Essential Fatty Acid Deficiency

This is rare in this country because of the use of Intralipid and may manifest as a scaly dermatitis. Should intravenous fats be contraindicated it may be prevented and treated by the cutaneous application of sunflower-seed oil (Press *et al.*, 1974).

PROLONGED INTRAVENOUS NUTRITION

Should the patient require total intravenous nutrition for more than three weeks, further advice should be sought. Prolonged intravenous feeding requires a great deal of skill and attention to detail. The catheter site has to be carefully selected, and a subcutaneous tunnel should be performed in order to reduce the incidence of catheter sepsis. A system may have to be evolved whereby the patient receives most of his feed at night via an infusion pump; this enables the patient to be mobile during the day, and under certain circumstances, it may be possible for the patient to be fed at home. Replacement of trace metals is essential.

REFERENCES

Ellis, B.W., Stanbridge, R. de L., Fielding, L.P. and Dudley, H.A.F. (1976). *Brit. Med. J.* 1, 1388.

Lee, H.A. and Hartley, T.F. (1975). *Post. Grad. Med. J.* 51, 441.

Press, M., Hartop, P.J. and Prottey, C. (1974). *Lancet* 1, 597.

SUBJECT INDEX

Pharmaceutical preparations are given in italics

R

S

T

U